Moral Injury and the Promise of Virtue

Joseph Wiinikka-Lydon

Moral Injury and the Promise of Virtue

palgrave
macmillan

Joseph Wiinikka-Lydon
Center for Ethics
University of Pardubice
Pardubice, Czech Republic

ISBN 978-3-030-32933-4   ISBN 978-3-030-32934-1 (eBook)
https://doi.org/10.1007/978-3-030-32934-1

Cover illustration: David Wiinikka-Lydon / Alamy Stock Photo

This Palgrave Macmillan imprint is published by the registered company Springer Nature Switzerland AG.
The registered company address is: Gewerbestrasse 11, 6330 Cham, Switzerland

The concepts of the virtues and the familiar words which name them are important since they help to make certain potentially nebulous areas of experience more open to inspection.
—Iris Murdoch, *The Sovereignty of Good*

For David

ACKNOWLEDGMENTS

This book began as a doctoral dissertation completed at Emory University's Graduate Division of Religion. I am grateful for the debts I have accumulated along the way and for those people to whom I am indebted. I have to thank, first, the staff, teachers, and colleagues at Emory's Graduate Division of Religion for possibly the best environment in which to grow as a writer and teacher. In particular, I would like to thank Bobbi Patterson, Tara Doyle, Don Seeman, Sara McClintock, and Timothy Jackson for listening to my questions and ideas and providing much-needed advice.

The best of Steven Tipton, Pam Hall, Ellen Ott Marshall, and Elizabeth Bounds are present in this work. Daily I reap the benefits of their caring mentorship. I am very lucky to have been their student.

I would also like to thank Emory graduates who, over the years, have provided feedback, friendship, and support, including Jesse Smith, James McCarthy, Matthew Pierce, Brian Powers, Amy DeBaets (and Brian), Georgette Ledgister, Sara Williams, Won Chul Shin, Sarah Muwahidah, Sarah MacDonald, Bradley Burroughs, Justin Latterell, Matt Tuininga, Kristyn Sessions, Carolyn Browning Helsel, Jake Myers, AnneMarie Mingo, Liz Whiting Pierce, and John Senior, among others, as the list has grown so long. In particular, I want to thank a group based in Atlanta whose sanity, brilliance, and love have meant everything to me. The best of Jennifer Ayres, Letitia Campbell, Jessica Vazquez Torres, Melissa Pagán, Elsie Barnhart, and Laura Cheifetz are at the heart of this work.

Beyond Emory, I have had the chance to share my thoughts with and receive feedback from students at the University of the South, Denison University, Birmingham-Southern College, and the University of Pardubice. I am a better writer and thinker because of them. My gratitude extends to colleagues at these institutions who have supported this work in various ways and who have extended their friendship, in particular, Sid Brown, Tam Parker, John Cort, Amy Cottrill, and M. Keely Sutton.

Members at the Center for Ethics at the University of Pardubice have helped me hone my ideas over the past 2 years, and I thank Niklas Forsberg, Kamila Pacovská, Ondřej Beran, Ondřej Krása, Matej Cíbik, Michael Campbell, Nora Hämäläinen, Hugo Strandberg, Tomaš Hejduk, Marina Barabas, Maria Balaska, Christopher Cordner, and Raimond Gaita, along with the support and patience of Tereza Mlejnková, Alena Kavičková, Tereza Pejšová, and Attila Bela Pato. I am also grateful for the support of the European Regional Development Fund and the state budget of the Czech Republic, which co-finances the Center for Ethics at the University of Pardubice, for allowing me the time to finalize this work.

A number of colleagues have been generous enough to read parts of this work in draft form, and I thank especially Aristotle Papanikolaou and Maria Antonaccio. I am also grateful for the help and patience of Amy Invernizzi at Palgrave Macmillan for helping to see this work through to its completion.

Ideas and sections of the present work were presented at research seminars at Uppsala University's Engaging Vulnerability project, as well as Åbo Akademie University's philosophy seminar. I thank those communities for their time and feedback. Parts of the present work have also been presented at conferences at the University of Vienna, the University of Pardubice, Sabanci University, the American Academy of Religion, and the Society of Christian Ethics, and I am grateful to those present who helped improve my arguments.

Most of all, I am grateful to the support of my family, including Timothy Lydon, Sharon Lydon, Kathy Valenti, and the extended Wiinikka and Nelson clan who made me one of their own. Everyone needs harbors to complete such a long labor, and they have been havens for me.

This book is dedicated to David, my husband, my closest friend, and my home.

*   *   *

Chapter 5 is an adapted version of an article that was originally published as Joseph Wiinikka-Lydon, "Mapping Moral Injury: Comparing Discourses of

Moral Harm." *Journal of Medicine and Philosophy* 44:2 (2019): 174–191. My thanks to Oxford University Press for allowing an adapted use of that article in these pages.

I gratefully acknowledge Taylor and Francis's permission to use a quote from Iris Murdoch's *The Sovereignty of Good* in the front matter.

CONTENTS

CHAPTER 1

Introduction

Zlatko Dizdarević was an editor of the Sarajevan daily newspaper *Oslobodjenje*, which was a critical news source for besieged Sarajevans during the war in Bosnia-Hercegovina at the end of the last century (1992–95). Writing in his wartime diary that would eventually be published as *Sarajevo: A War Journal,* Dizdarević reflected on the way that the war's atrocities attacked what might be thought of as social virtues fundamental to the cohesion of community.[1] He writes, "The rifle butt in the back, and the truck ride to the camp cause a distress that cannot be forgotten. That rifle butt shatters everything civilization has ever accomplished, removes all finer human sentiments, and wipes out any sense of justice, compassion, and forgiveness."[2]

Although specific to the Bosnian conflict, this striking quotation also captures a more general experience of political violence that others, in other wars, share. Dizdarević bears witness to the experience of those who survive political violence—such as genocide, long-term siege, and intense violent conflict—only to discover that they and their world, or at least their moral perception of that world, have changed in important, fundamental ways. In other words, Dizdarević is an example of those who survive mass violence yet, because of what they have witnessed or what they

---

[1] For *Oslobodjenje's* role in Sarajevo's resistance to the siege, see Kurspahi, *As Long as Sarajevo Exists.* Dizdarević's diary was eventually published as a record of his work and reflections during the multi-year siege. See Dizdarević, *Sarajevo: A War Journal.*

[2] Dizdarević, *Sarajevo,* 54.

© The Author(s) 2019
J. Wiinikka-Lydon, *Moral Injury and the Promise of Virtue,*
https://doi.org/10.1007/978-3-030-32934-1_1

may have done to survive, are left with the sense that they or the world have changed in a profound moral sense. Survivors may come out of the conflict feeling they have done so only by committing acts previously thought to be immoral or by witnessing events, such as the one Dizdarević reflects upon, that shake one's faith in the very possibility of goodness in such a world. This can leave a lasting change to one's moral subjectivity, so much so that survivors often wonder if they can ever be "good" again or what the worth of trying to be good is in such a fallen world.

Sarajevo's was one of the longest sieges in modern military history that included daily fear of random sniper and mortar fire. This created an environment of deep uncertainty and anxiety that wore away the daily patterns of living and identity that existed prior to the war. The war itself ended with over 200,000 killed, the creation of concentration camps on European soil for the first time since World War II, and the "ethnic cleansing" of the country, a term that originated during the conflict to refer to the forced elimination of a group, their history, and their cultural artifacts from a certain area. It was a war, a genocide, whose numbers seem small when compared to the rapid carnage of the Rwandan genocide occurring at the same time. Yet, a profound aspect of the war in Bosnia-Hercegovina, and in the former Yugoslavia more generally, was the destruction not just of individual lives but also of the culture and ties that enabled community.

What the testimonies of those like Dizdarević suggest is that through one's actions or the actions of others, political violence can undermine the moral intelligibility of one's world. Violent conflict does not just destroy lives and buildings. It destroys that which makes meaningful lives and community possible. Experiences of war like these can leave a residue of doubt and even despair about the possibility of a restored moral ability and a world capable of supporting a meaningful moral life. It is, in other words, a moral harm that has a lasting impact on how one perceives oneself and others, and on the ability for one to actively aspire to the visions of goodness and the images of personhood that are central to the moral dimension of life, and so are central to one's existence.

Unseen though they may be, such changes have profound consequences for society. As these images and visions of self, world, and other tell us who we are and who we should be, they are also central to how people understand themselves. They are central, then, to identity, and so changes to one's moral architecture will also change one's identity. These moral harms, what have been called in the context of veterans, "moral injuries," are significant, then, not only for individuals but also for whole societies. Since these harms affect identity, they will also affect society, and if powerful enough, will transform society as well.

Such experiences raise serious questions not only for survivors of political violence but also more broadly. What happens when violence changes an individual's moral architecture so dramatically? What is it that is affected? And how can we talk about such moral transformation in a way that can illuminate the relationships between violence and subjectivity, of society and the individual, and our understanding of moral experience and character?

Attempts to investigate such questions, which deal with the effects of violence on subjectivity, have largely come from the social sciences. Entire subfields in sociology, anthropology, and social psychology have investigated the causes and dynamics of violence from various angles. It would therefore be reasonable to look there for methods and approaches that could help make sense of the experience that Dizdarević and others have faced. Even with the significant insights such research has provided, however, there is a curious absence of strongly normative language to more fully articulate and account for the experience that survivors have claimed is so central to what is at stake in violent conflict and war. Sociologist Andrew Sayer, for example, argues from within the social sciences that these fields seem to be missing strong frames and vocabularies for representing such claims in their research. As his critique comes from within these fields, it is worth quoting Sayer at length in this regard:

> … we are beings whose relation to the world is one of concern. Yet social science often ignores this relation and hence fails to acknowledge what is most important to people. Concepts such as "preferences," "self-interest," and "values" fail to do justice to such matters, particularly with regard to their social character and connection to events and social relations, and their emotional force. Similarly, concepts such as convention, habit, discourses, socialization, reciprocity, exchange, discipline, power, and a host of others are useful for external description but can easily allow us to miss people's first-person evaluative relation to the world and the force of their evaluations. When social science disregards this concern, as if it were merely an incidental, subjective accompaniment to what happens, it can produce an alienated and alienating view of social life.[3]

---

[3] Sayer, *Why Things Matter to People*, 2.

Every method, every discipline, has its limits, but what Sayer is arguing is that there are unnecessary restrictions, or perhaps it is better to say there is a general reluctance to examine social life through the lens and with the vocabulary of normative concepts and vocabulary. Denying such an approach leaves out the eye level, that is, any subjective account of humans living their everyday lives that takes personal evaluations into account. This runs the danger that any analysis of social structures, institutions, or actions will not only be limited but also be critically incomplete.

This lacuna is even more interesting if we consider that concern over mental and moral harms have only increased over the years. The last several decades have seen an expansion of studies dedicated to violence, both structural and episodic, as well as to the ways in which violence can transform individuals in negative ways.[4] Terms such as *psychological trauma, post-traumatic stress disorder*, and, more recently, *moral injury* signal the creation of a vocabulary reflecting experiences that, while undoubtedly quite old, are nevertheless receiving concerted attention from government, civil society, and academic inquiry. Changes in terminology regarding psychological combat trauma, for example, in the change from *a soldier's heart* and *combat hysteria* in the nineteenth century to *war neurosis* and *shell shock* before and after World War I, respectively, to *post-traumatic stress disorder* during the Vietnam War and after, evince, if nothing else, cultural debates over the effects of policies and narratives on an individual's ability to cope as a peaceful moral subject within society, once the context of explicit violence, such as war, has officially ended.[5]

My reason for highlighting this lacuna is not to delve into the history of the academic disciplines, nor to dismiss the research of decades and even centuries. It is, instead, to highlight a need on the methodological level in disciplines that have proven invaluable for explicating social, cultural, and psychological dynamics, particularly where it concerns social upheaval and political violence. This is important for three main reasons.

[4] For example, the very term "structural violence" was only coined in 1969 and studies into the violence inherent in institutionality found in Foucault and Bourdieu occurred only in the second half of the twentieth century (Galtung, "Violence, Peace, and Peace Research"). Post-traumatic stress disorder (PTSD), which has almost become a colloquial designation, was only added to the American Psychiatric Association's *Diagnostic and Statistical Manual of Mental Disorders (DSM)* in 1980 (Trimble, "Post-Traumatic Stress Disorder," 5, 12).

[5] Trimble, "Post-Traumatic Stress Disorder," 5–7; Crocq, "From Shell Shock and War Neurosis to PTSD."

First, if we do not have frames that privilege the moral dimension of violence, we run the risk of privileging other aspects of violent conflict and obscuring the moral. This can cause researchers to miss an important aspect of human experience, one that not only motivates individuals and groups but is also central to meaning and the experience of being a self in relationship to others, however that may be defined. Although he did not focus on violent conflict, these are also Sawyer's concerns, and they have been seconded most recently in anthropology by those like James Laidlaw.[6] Laidlaw worries that social and cultural analysis overly emphasizes deterministic elements of human life at the expense of the experience of individuals as moral agents in the world. In this view, the social sciences too easily subsume the ethical and ethical life into the "social." Laidlaw's is not an argument to make of social science a virtue ethic, as he states such approaches to do not make an "evaluative claim that people are good: it is a descriptive claim that they are evaluative."[7] Instead, he is focused on bringing the discourse of "freedom" back into the heart of anthropology, a move that, although pushing back on overly deterministic approaches, may itself reveal a "Western" bias that Laidlaw himself is trying to prevent.[8]

As I will illustrate in the following chapters, context seems to overwhelm the experience of Dizdarević and Maček's informants. They, as well as others I present, articulate the experience of being forced into certain identities and banned from others. Although I agree with much of Laidlaw's work, including his focus, which this book shares, on moral development (making it in part a "pedagogical ethic"), the ways in which violence can overwhelm identity makes discussion of freedom and Laidlaw's focus on ethics as, following Foucault, a "technique of self-formation," problematic for understanding such survivors, except perhaps in freedom's negation. There remains, then, beyond these debates that concern the proper nature of ethnography or social research, a practical necessity. Specifically, it is the need of those whose experience of violence

---

[6] Sociologist Philip Gorski has also written on Emile Durkheim's use of Aristotle to show the relevance of virtue ethics to sociology (Gorski, "Recovered Goods"). And anthropologist Thomas Widlock has also argued for virtue as a salient category in anthropology (Widlock, "Virtue" and "Sharing by Default?") Their work is different enough from the present study to include only a mention here, but they are part of a growing trend to recover moral concepts for the social sciences, although for a varied range of reasons.

[7] Laidlaw, *The Subject of Virtue*, 3.

[8] Mattingly, "Two virtue ethics," 162; Robbins, "Between reproduction and freedom," 295.

is so profound that such experiences seem not only to resist articulation but may even disintegrate the possibility of language on the very tip of one's tongue.[9] It is a need that, although connected to social-scientific discussions of the proper place of ethics within these disciplines, exceeds these works. The experience of survivors of such violence—the violence of war, genocide, and siege that includes torture, starvation, rape, and profound moral challenge—requires a language, and really an interpretive framework in which they can make sense of their experience, that connects to the insights of the social sciences and the work of those like Laidlaw, Sawyer, and others. It must, however, go beyond them to be a frame that survivors and researchers can use to provide a language in situations where our everyday language seems to fail.[10]

This leads to the second reason for highlighting this lacuna. If there is no vocabulary able to reflect this moral dimension of survivors' experiences, there is the danger that any research into such conflict will not have the resources even to frame the issue in moral terms. Such a lack makes it much more likely that any moral stakes will be misconstrued, subsumed into other analytical categories like the social or the economic, or overlooked.

This has a direct effect on my third point, that without vocabularies to reflect such experience, and without frames that take into account the moral dimension of social life, the testimony of survivors will be elided. I am talking here about a methodological issue but one that is not merely academic. The social sciences are important. They are an influential field in which knowledge is produced, knowledge that is often used to influence policy. That the voices of survivors of political violence are elided, first on the methodological level and then in a second consequential sense, on the level of the application of research findings that are based on such methodologies, is more than a disciplinary matter. It is also a justice concern. Those made vulnerable through violence can be made vulnerable again, not only through methodological and analytical elision but also

---

[9] Scarry, *The Body in Pain*.

[10] As the reader will soon see, I am assuming the experience of survivors from the war in Bosnia-Hercegovina in the 1990s. What counts as "everyday language," and what is called "ordinary ethics" by Veena Das ("Ordinary Ethics") and Michael Lambek ("Toward an Ethics of the Act") will depend on the history of a given place and people, some of which have a great deal of experience with extreme violence. The insights gleaned from this work should be a resource for such situations, though it will require a critical engagement based on the experience of those living in the midst of such violence.

through public policies that repeat this elision on the level of application, as such policies effect political recognition, distributive justice, and access to the civil and political spheres.

At issue are testimonies concerning the ways war and political violence can change individuals and societies on a moral level. What I mean is that such violence can, through direct violence on persons, as well as through the transformation of institutions and public space where individuals interact, transform what types of identity and society one can imagine and inhabit. This is a central aspect to political violence that affects not only moral development but also social futures, as the ability to imagine different futures directly affects post-war peacebuilding efforts and their possibilities of success. Seemingly intangible, morality, moral imagination, and moral development are deeply important aspects of material intervention into social futures.

Without appropriate terms and images that reflect the above criteria, then, researchers will try to hit a target with the wrong instrument and may not even appreciate the target's existence. Would a sociologist, for example, be expected to account for the rise of nongovernmental organizations without a vocabulary of institutions, politics, and social development? Would we expect a medical researcher to account for changes in physiology without a rich, specialized vocabulary that arises from many centuries of medical discussion, debate, inquiry, and refinement that was focused on the body? These examples illustrate that it makes no sense to investigate issues of moral concern—of concern to individuals and their intimate relationships—without a vocabulary and understanding of the self-designed to speak to such a dimension of experience and social life. Without the availability of such frames, an aspect of violence and subjectivity that survivors claim is central is in danger of being marginalized, devalued, or elided in the midst of good faith work.

We are dealing, then, with difficult matters, a difficulty that is felt by both researchers and informants. What we require, then, is an understanding of the individual as primarily a moral subject for whom events and experience are first and foremost moral matters, that is, that their impact is felt in that they affect one's worldview and moral development. The vocabulary needed, then, needs to be capable of reflecting such experience and of articulating the meaning of experience in moral terms. "Moral" here means not just issues of right and wrong but, as Charles Taylor might articulate it, more robustly as an orientation to what makes life worth living. I would expand on this somewhat and describe "moral" as an orientation toward goodness, however that is understood, and the development that enables one to live a life aimed at goodness, embodied in particular images of the human and other horizons, and directed away from the vicious and vicious, unvirtuous images of the human. It reflects, as Bernard Williams has said about Socrates, the primary concern of individuals over who they are, who they want to be, and how to be in the world and toward others. It is an understanding of the moral dimension of experience, what is often called the moral life, that is not reduced to a code or rules but focuses on orientation, images and symbols, narratives, horizons, gestures, meaning, and embodied practice that, importantly, make life into continuous moral development, creating a constantly changing life and one's understanding of their life and its meaning.

How to Speak of Moral Change

What vocabulary, what approaches, can fill this need? If we return to Dizdarević's statement, he has, in a way, indicated a possible way forward, one that the rest of this book will explore. His choice of language, referring to finer sentiments, of humanity, and of certain virtues gestures toward virtue languages and in general to that mode of ethical reason and practice usually referred to as virtue ethics or virtue theories. Dizdarević argued that violence during the war in Bosnia destroyed the very values or practices, such as *compassion, forgiveness*, and the possibility of *justice*, that make social life possible. Compassion, forgiveness, and human relations are not only characterological but also, it can be argued, more specifically virtues and dispositions that, in the classical understanding of virtue, help one reach toward certain excellences of understanding, of action, of relationship, and of community. These are elements of social cohesion that, unlike bricks and mortar, are not so easily replaced, and in his effort to articulate such loss, Dizdarević turns to the language of virtue to describe this loss and not to economic or political-scientific categories, although as a celebrated journalist and editor, he was certainly aware of such discourses.[11] He uses a colloquial virtue language to attempt to communi-

[11] This, of course, does not mean that such experiences do not have economic causes at their root, as there was a great deal of economic decline that helped precipitate the conflicts. Indeed, much has been written on this conflict and on the related conflicts associated with the dissolution of Yugoslavia, focusing on the economic aspects of the war and its causes, while other examples of such research have looked at the history and politics of the region. Perica, *Balkan Idols*, 308–9; Ramet, *Social Currents in Eastern Europe*, 374, 414. A selection

cate the profound loss that has occurred in the war, a loss occurring on the intimate interpersonal level, one that can all too easily be missed by too much attention being paid to macro social, economic, and political events and dynamics.

Instead, Dizdarević decries the events he witnessed in *moral* terms, using virtue language to underline the moral dimension of what was at stake during the war. Although more will need to be said about how *"moral"* will be understood in these pages, for the moment we can understand *"moral"* in developmental terms as the ability to live into or reach toward normative, even prescriptive, horizons of behavior and relationality that provide identity and meaning to one's life shared in common. Using such language, Dizdarević argues that the ability to be a person that can embody that vision of *civilization* has been destroyed. A way of being—and a particular understanding of how to be a human being—has been lost. What we have been deprived of, then, are virtues and qualities of character that allow one to become a morally realized person, perhaps even to aspire to an ideal way of being human, what we could surmise from his short passage to be a civilized man or woman. To be more exact, Dizdarević can be read as referring to the felt experience of having lost such capacities, capacities that empower one to be able to flourish within the potential horizon of a civilized world. This resonates with classical understandings of virtue that are *eudaimonistic*, whereby virtues enable one to live and act in a way that can consequently support a flourishing life.

Dizdarević speaks of what is at stake particularly in terms of "justice, compassion, and forgiveness," in other words, prosocial virtues that were harmed, even destroyed, during the course of the war. When seen as traits, these are virtues of a good neighbor that enable trust and cooperation within a community and without which community falls apart. The very possibility of these virtues—and, when seen as shared goals, of values— are, then, central to shared life. To claim their destruction is to claim that community is no longer possible, and in a sense, then, morality is no longer possible. Something has happened to the world through the violent

includes: Malcolm, *Bosnia: A History*; Brubaker, *Nationalism Reframed*; Waller, *Becoming Evil*; Glenny, *The Fall of Yugoslavia*; Vulliamy, *Seasons in Hell*; Perica, *Balkan Idols*; Ali and Lifschultz (eds.), *Why Bosnia?*; Ramet, *Balkan Babel*; Ramet, *Social Currents in Eastern Europe*; Denitch, *Ethnic Nationalism*; Banac, *The National Question in Yugoslavia*; Norman, *Genocide in Bosnia*; Mojzes (ed.), *Religion and the War in Bosnia*; Sells, *The Bridge Betrayed*; Allen, *Rape Warfare*; Campbell, "Violence, Justice, and Identity in the Bosnian Conflict." See also Brubaker 1996, 64–65, 70, 73.

actions of others that no longer allows the pursuit of goodness. It is a metaphysical crisis. Such situations can create the impression of betrayal and also unintelligibility, as a world no longer capable of goodness is an alien world where one is no longer sure what horizons are worthy of one's effort. Indeed, without the possibility of such virtues and horizons toward which one can aspire, it is a world that may no longer exert a pull on one to improve. In its place are anxiety, lack of trust, resentment, and even despair.

Not only descriptive and normative, the virtue language he uses is also critical and prescriptive. Such vocabulary does not just describe, but also, because of its prescriptive sense, it is able to make claims, to critically analyze situations, and even to condemn. Such a vocabulary can illustrate uniquely and powerfully the moral stakes involved, and in so doing indicts the conflict and those who propagate it. This indictment Dizdarević aims at those perpetrating the conflict for destroying cultural and social achievements that make a certain quality of life, which involves "finer sentiments" such as "compassion" and "forgiveness," impossible. Without these finer qualities, people themselves are less fine, perhaps coarser, less able to reach for "civilization" and the horizons of cultural and even spiritual realization that such a heavily normative term implies. The teleological possibilities of human individuals, the shared human life we call the social, and the ways that enable such teleologies that we call culture are themselves circumscribed, cutting off important ways that we can be fully human. The indictment, then, is clear: what is at stake is humanity itself. And what survivors are left feeling, possibly, is less human, and possibly set adrift in a world less capable of giving rise to humanity.

Dizdarević is not alone in his concern. Ivana Maček, who conducted fieldwork in Sarajevo during the war in the 1990s, also argued that one of the main anxieties of those caught in Sarajevo during the siege concerned not just safety for themselves and their loved ones but also the ways in which violence and the conditions imposed by the war were transforming their moral being. Maček writes:

> The preoccupation with normality reflected Sarajevans' utmost fear and their utmost shame: that in coping with the inhumane conditions of war, they had also become dehumanized and that they might be surviving only by means they would previously have rejected as immoral. Had they become psychologically, socially, and culturally unfit to live among decent people?

What Maček heard in the midst of the violence was a concern not just over survival, as important as that was, but also over the quality of that survival. There was a fear among her informants that they were losing touch with who they once were. They were changing, and in doing so, they were losing contact with the ability not just to live in a certain way but also to be a certain type of person. Indeed, Maček is not the only one to encounter such expressions. Journalist Barbara Demick, who also lived in Sarajevo for large stretches during the war, reported that people she met were very eager to show her pictures of vacations in Europe and the nice clothes they had. They were intent on showing that they had not always lived holed away and scrounging for their daily existence. They wanted to demonstrate that how they appeared during the war, and how they acted, was not who they really were.[12]

This concern, expressed in terms of anxiety with *normality*, articulates the loss of a world, a world in which certain cherished norms and ideals and visions of the human are possible. In the midst of violence, one of the key stakes was not just survival but also the survival of the possibility of living into certain norms and visions of the human being—in other words, one's own survival as a being capable of being moral in a certain way. Some of Maček's informants called the *imitation of life* the experience of living under siege, how it seemed so strange from the vantage point of their pre-war lives as to be not real but a simulacrum of a once-experienced reality. This loss of normality did not just refer to the loss of employment or even residence, however. It also represented a loss in moral ability and moral status as a good person and citizen, something that Maček is keen to recognize and explain. *Normal* carried a "moral charge" and "pertained not only to the way of life people felt they had lost but also to a moral framework that might guide their actions. Normality not only communicated the social norms held by the person using it but also indicated her or his ideological position."

What both Maček and Dizdarević gesture toward is a need for moral terminology not only to describe what has been lost during the war but also to help explain how such loss occurs. Although Dizdarević uses virtue terminology, however, neither he nor Maček have a sufficiently rich, systematized frame to fully make this possible. Dizdarević himself does not have a sufficient vocabulary to communicate the danger he saw, using words such as "civilization" with their inexact, but even more importantly,

---

[12] Demick, *Lugovina Street*, xxxi.

violent histories. This makes sense. How many of us learn a vocabulary for such devastation? How many take the possibility seriously that we may experience something similar? Who prepares words for the collapse of one's worldview and felt ability to strive after being good, however we define that?

Such a need is not presaged but only acknowledged in its aftermath. This makes the words of those such as Dizdarević all the more important, as they unearth the possibility for such a need, even as the difficulty in articulating those stakes are also revealed in that exhumation. Maček's account also suffers from a lack of moral language that could not only give full voice to these experiences but also account for what exactly changed for individuals. She is no doubt right that violence does disturb norms, as it can also sharpen the need to fight for them. And she describes brilliantly the precariousness that such violence creates in the foundations of one's faith in the moral coherence of the world, as well as in its potential for meaning. The idea of *normality*, however, which is a deeply moral term, is not fully fleshed out. There is no morally rich vocabulary, and no understanding of the human that prioritizes the moral dimension of experience and individual development, to explain sufficiently how exactly this comes to be and what exactly about the individual changes in situations of political violence.[13]

---

[13] Maček makes this search of language for the experience of the war central to her work, which she frames in terms of "limit situations" (Maček, *Sarajevo Under Siege*, 30). To provide a more theoretical explication for her informant's experiences, Maček brings to it terms from Holocaust research and writings, such as Primo Levi's gray zone, and talks about the humiliation of having no power over one's life (Maček, *Sarajevo Under Siege*, 66–70). She also documents the ways in which the besieged resisted indignities by trying to hold on to the aesthetics and routines of pre-war life, through the use of humor, and the creation of pride by overcoming wartime obstacles. And Maček compiles this evidence to argue for violence as the key source of the dissolution of ethical systems as opposed to, say, competition (Maček, *Sarajevo Under Siege*, 6–7, 88–89). But this method does not include a further frame based on a moral language to explicate the moral dimensions of *normality*, which is an inherently moral term.

Although vocabularies are limited, there are nevertheless various works trying to explicate the experience of violence. Some of the more well-known include Elaine Scarry's *The Body in Pain*, Judith Herman's *Trauma and Recovery*, Susan Brison's *Aftermath*, and more recently J.M. Bernstein's *Torture and Dignity* that draws not only on Brison's work but also on Jean Améry's classic, *At the Mind's Limits*. There is also now a growing literature on the experience of moral injury, understood as harm done to veterans by their role as soldiers, such as Gabriella Lettini and Rita Brock's *Soul Repair*.

Argument for Virtue Language

This reading is not intended to put words in the mouth of the author of *Sarajevo: A War Journal*. The intention, instead, is to use his reflections, and the insights of Maček's ethnography, as a way to help a broader group of people better understand those experiences that he tries to describe. It is an occasion, one that I argue would be profitably spent looking into ways that languages of virtue could assist researchers, as well as those who experience such violence, to better articulate and account for the form of harm to which Dizdarević and Maček attest.

In the following pages, I develop the possibilities of the language of virtue to better articulate and account for the type of loss that people who have survived violent conflict have tried to give voice. What I strive to demonstrate in this work is how philosophical and religious ethics can provide the vocabulary that we need to help fill this lacuna. Virtue discourse, more so than current approaches in political violence and subjectivity, can better capture the nature of selves and their formation, thus accounting for how violence can unmake an individual's moral architecture. The many writers and traditions that fit under its umbrella have developed rich descriptions of the human, of morality and moral development, and of the social and even institutional contexts in which the moral life is lived. This history and its conceptual resources can provide a firm basis on which to develop an understanding of subjectivity that privileges the experience of the individual as a moral subject and of society and culture as domains where value and moral subjects are made, maintained, and transformed. In other words, virtue language and theories can provide a basis for a vocabulary and conception of the self that can best account for and articulate the experience of political violence captured in such testimony as that of Dizdarević, as well as others that we will see in the following pages.[14]

---

[14] This does not mean, however, that other moral languages and modalities are not useful. There may be good reasons why a neo-Kantian or utilitarian vocabulary could be developed and employed, particularly if one's goals differ from that of the present inquiry. To try and compare and contrast different moral-philosophical languages would take a volume in itself, however. The goal of this work is not to arbitrate between these options but, instead, to use a specific resource—virtue—to show that virtue can be applied in a critical, analytical way to questions of violence and subjectivity in a way that current methods of violence and subjectivity do not. Comparisons between a virtue hermeneutic of moral subjectivity and other hermeneutics are welcome but will need to be saved for future works and conversations.

There are several specific reasons I can delineate for engaging the language of the virtues. From its beginning in southeastern Europe, virtue ethics, at least in the form derived from ancient Attic philosophy, sought to explain the point of human life in constant reference to the goal of human life that is shared, that is, politics.[15] Although Aristotle did not have the language of modern institutional analysis, he famously understood that humans are *political* animals, and explored in his *Politics* institutions such as the family, the state, slavery, as well as others, as the contexts within which we develop morally. Plato was also highly concerned with political relationships and institutions, dedicating two of the most influential works in philosophy, *The Republic* and *The Laws*, to discussions of moral development, institutionality, and political philosophy. Although virtue ethics is best understood as an umbrella term for a great variety of thought concerning ethics and moral development, there remains in certain modern thinkers, most notably Alasdair MacIntyre and, in a different way, Martha Nussbaum, a focus on the communal and the political.[16]

In addition to this attention to the political and the social context of one's moral development, virtue ethics also gives us a philosophical anthropology that can emphasize moral development when speaking of the self. For example, Elizabeth Anscombe's call for renewal in moral philosophy at the middle of the twentieth century, a call now seen as the beginning of a attention to virtue in modern philosophy, specifically argued for the development of new moral psychologies.[17] These psychologies account for how one is to develop morally, what one needs to do so, and how the individual is constituted. This constitution includes emotion, purpose, cognition, which provides an account not just of *what*

[15] Scholars over the last few decades have argued that virtue ethics arose in other locations, arguing for virtue ethics in Buddhism and Confucianism, to use two examples (Tu, *Confucian Thought*; McKeown, "Buddhism and Ecology;" *The Nature of Buddhist Ethics*, Bary and Tu, *Confucianism and Human Rights*). I will focus, however, on virtue ethics in so-called Western philosophy, meaning those understandings of virtue rooted in the philosophy of Attic Greek nation states in the last millennium "Before the Common Era."

[16] MacIntyre, *After Virtue*. Nussbaum is more difficult to discuss in terms of virtue ethics, as she sees to have created a critical distance between her thought and that category, moving to develop instead, along with Amartya Sen, a capabilities approach grounded in a liberal, Rawlsian political philosophy (Nussbaum, *Frontiers of Justice*; *Women and Human Development*). She has contributed to virtue ethics, however, and has an eye for how virtue affects life and institutions.

[17] Anscombe, "Modern Moral Philosophy," 1.

morality and virtue are but also of *how* one is to be and become virtuous. Such theories usually assume or reference, if not describe, practices necessary for moral development along with related ends and goals. By showing what is necessary for moral development, they also show in what ways human moral trajectories are vulnerable. This is critical for any study such as this one, which aims to show how individuals feel moral loss through violence.[18]

Virtue discourses, then, supply us with a metaphysics, a psychology, and a sociology that are bound up with political theories and understandings of practices to account for the development of the human person. They look at human lives as robust, inquiring into the effects on moral development of art, music, poetry, sport, social organization, and political systems. This includes not only cognition but also imagination and emotion, and even understandings of embodiment are included, or, at the very least, exist as potentials. Virtue discourses provide then, resources for frames that can account for individuals socially and culturally embedded, as well as embodied, providing a very robust, detailed representation of the human.

The vocabulary used in virtue discourse is also important, as it is what we might call *morally saturated* or *invested*. The terms used in virtue language were created over time to expound upon and help answer questions about human life, questions of meaning, how one should act in the world, and understandings of the best and best possible societies. These are very normative issues, moral issues that, in Sayer's words, concern that which matters most to people, what makes life worth living, what contributes to a good life shared in common, and what is necessary developmentally on the individual and social level for this to occur. Such language, of course, is not neutral; yet this is the point. It was developed in the process of prescribing certain ways of life and how to grow as a "good" person, and so has the conceptual sophistication to describe related issues.

Although one could argue that colloquial uses of virtue language are more common, I appeal to such systematic ethics to draw out the language and descriptive scaffolding used to construct their ethics. This is not meant to privilege the formal over the everyday or the academy over the street. Those who have most thoroughly systematized virtue discourse are those involved in moral philosophy. Here we can find more readily avail-

---

[18] cf., Williams, "Moral Luck."

able languages. I look to virtue ethics as well, due to the goals of this study. A major goal is to demonstrate that virtue can be used in creating an interpretive framework that can make better sense of an experience of violence, subjectivity, and change that survivors have said is important. It is important to investigate the different moral discourses that people use when coming from contexts other than the orthodox, elite, or scholarly. Indeed, it is an exciting prospect to think of such a discourse being articulated and then used to articulate and account for experiences found in that locale. Yet, there is still something to be said for including more formal accounts, as they too have had broad influence. More importantly, part of my argument is that we can adopt moral languages and employ them as a critical interpretive frame used in other ethnographic projects and related inquiries. In developing my argument here, my focus is to show that such a frame can be applied successfully, which does not require the development of an ethnographically derived discourse. Although the ethnographic method I argue would be important in diversifying our understanding of virtue discourse by including cultures, communities, concepts, and voices not presently included in formal academic ethics, such work will have to wait for a future work.

In summary, then, there are four main reasons I turn to virtue discourse as developed through the history of virtue ethics to explore the effects of violence on subjectivity. First, such language draws from centuries of philosophical discussion concerning the nature and ends of the human being. This represents, then, a rich reservoir from which we can draw to discuss and analyze the moral dimensions of social and political experience. Further, virtue theories and ethics presuppose a political and metaphysical background that includes accounts of how the self relates to society and the broader world.[19] The focus is on the individual, then, but specifically the individual who is fully imbedded in the social, cultural, and political and one who is acting constantly within culture and social institutions as a norm-making, norm-contesting, and norm-following being—a moral

---

[19] Although I am not engaging with metaphysical questions here, it is safe to say that metaphysics has been critical for most traditional virtue ethics, although a biology to ground one's understanding of virtue can also be emphasized, such as in the ethics of Aristotle, who developed not only personal ethics but also political philosophies and biologies. One of the most prominent neo-Aristotelians (and Thomists, for that matter), Alasdair MacIntyre, wrote *Dependent Rational Animals* after *After Virtue* in order to provide a missing biological grounding for his moral philosophy (MacIntyre, *Dependent Rational Animals*, x).

subject. Third, theories of virtue ethics have a philosophical anthropology. They account for how one is to develop morally, what one needs to do so, and how the individual is constituted, including emotion, purpose, and cognition. This provides an account not just of what morality and virtue are but also of how one is to be and become virtuous. A fourth and related reason is that such theories usually assume, if not describe, practices necessary for moral development along with related ends and goals. Such discussions of development often include an account of how vulnerable one's moral being is and the ways it may be undermined. All of this is important for this study, which focuses on the individual transformed through political and social upheaval.

I have argued that virtue language has a strong descriptive capacity, owing to the need of virtue ethicists over the centuries to account in detail for the elements that make up the moral life and society seen as a moral body. It can supply a way for persons to analyze particular phenomena, as well as to articulate experience and to name what was at stake, what was lost, and what has to be gained. It should, optimally, provide a way for survivors or others involved in political violence to more precisely speak of their experience, and for such subjects to advocate for their own understanding of the good that has been and now is in their lives. The fact that such vocabulary is normative from the perspective of the subjects, as well as descriptive, means that it can provide a fuller account that captures the normative dimension of such loss, captures the stakes involved, and so helps give those who have survived words to articulate difficult thoughts and feelings.

Further, this work contributes to two other conversations. The first is a conversation between those interested in virtue discourse and theory. Quite often, virtue is discussed as a characteristic or disposition that one has, and so quite a lot of writing that focuses on virtue is concerned with specific virtues, how virtues are inculcated, or whether or not virtues are real. Since Anscombe's seminal "Modern Moral Philosophy," there has been the development of virtue ethics as mainly a counter-normative approach to utilitarianism, social contract theory, and neo-Kantian and deontological theories. Anscombe's article is often given credit for the explosion of this new mode of ethics, which mostly focuses on normative debates within modern academic, particularly analytic, philosophy. Drawing largely, though not exclusively, on neo-Aristotelian thought, this work has expanded the conversation within academic philosophy concerning what normative frameworks are possible in accounting for moral reasoning and action, epistemology (through virtue epistemologies), and practical thought.[20]

Although such projects are not unimportant, this work argues, in part, that they do not exhaust the potential of virtue. The rich, descriptive language found in virtue theories, as well as the understanding of human being often developed in virtue approaches, has the potential as an interpretive frame to provide insight into cultural and social dynamics and events. For example, philosopher Lisa Tessman, working from a neo-Aristotelian framework, argues that virtue—what she calls "critical virtue ethics"—can be used to critique social injustice and better account for the ways in which oppression and work toward justice can cause moral harm. Christine Swanton, who has been influenced largely by Nietzsche in her understanding of virtue, argues from separate sources that we need to conceive of flourishing in a way that includes a wide variety of lives, such as those referred to as "disabled." Swanton and Tessman both illustrate ways in which virtue ethics and, more broadly, the language used to discuss virtue, can help us better interpret modern social challenges and relationships.[21]

The second conversation this work contributes to concerns those discussions in anthropology involving subjectivity and violence. I am thinking here of anthropologists such as Arthur and Joan Kleinman, Veena Das, Begoña Aretxaga, Nancy Scheper-Hughes, and Philippe Bourgois, among others. What these researchers share is an emphasis on the embodied individual as the subject of research, a subject impacted by violence, though one already shot through with social structure and histories and cultural narratives. They take seriously the embedded nature of the embodied self as a nexus of complex events and forces that make up who we are from moment to moment and day to day. These writers also, however, make the connections between the transformation of the individual and society, and vice versa. They aim to capture and account for this dynamism yet do so

---

[20] This literature is quite large and diverse, covering not only moral virtues, which are the focus of this work, but also epistemic virtues in order to try to solve problems in traditional epistemology (Zagzebski, *Virtues of the Mind*). There are also various sources for such ethics. Although Aristotle has loomed large, Aquinas, Hume, Nietzsche, Martineau, Hutchinson, and Plato have also been resources. There have even been attempts to meld virtue with other normative approaches, such as a utilitarian virtue ethic (Driver, *Uneasy Virtue*). For good overviews of current and seminal works in virtue ethics, see Crisp and Slote, *Virtue Ethics*, which is a classic source.

[21] Tessman, "Critical Virtue Ethics" and *Burdened Virtues*; Swanton, *Virtue Ethics*.

by privileging the local level of the individual and community to capture the experience of those suffering from violence. It is an approach that is both engaged and humane. This allows them to take seriously how radical changes to culture, society, and politics through violence can alter one's very makeup through the alteration of the makeup of one's surroundings.

This work on subjectivity and violence is highly insightful and has been a main inspiration for the work included in this book. Even so, such research can still lack a vocabulary able to capture adequately the experience, even the interiority, of what it is to be changed in one's moral being, that is, to feel that one's moral architecture has been changed, weakened, or undone. The work of Arthur Kleinman, for example, has been particularly seminal in work on violence and subjectivity, both in bringing together likeminded researchers as well as developing such concepts as the *local*, concepts that we will engage constructively in the following chapters.[22] For Kleinman, humans live on the local level of social existence, but, as we will see, this local level is not just geographic but is also generatively normative. He refers specifically to *local moral worlds* to stress that it is within these networks of relationships where our ideals and moral worldviews are developed. It is this phenomenological level where experience occurs, making for Kleinman a necessary connection between experience and the moral. Indeed, it is not just a connection, but also an intertwining and even an understanding of the local as the place where experience (understood as the subjective, embodied [felt] interaction with and between self and world, mediated through already existing memories, histories, cultural narratives, and social structures that creates an always-transforming and transformative worldview and ethos) and moral codetermine one another.[23] As Kleinman argues, "Experience (including its sociosomatic interconnections) is innately moral, because it is in local worlds that the relational elements of social existence in which people have the greatest stake are played out."[24]

---

[22] Das et al., *Remaking a World*; Das et al., *Violence and Subjectivity*; Kleinman et al., *Social Suffering*; Kleinman et al., *Deep China*.

[23] This is similar to Kleinman's own definition: "Experience may, on theoretical grounds, be thought of as the intersubjective medium of social interactions in local moral worlds. It is the outcome of cultural categories and social structures interacting with the psychophysiological process such that a mediating world is constituted. Experience is the felt flow of that intersubjective medium" (Kleinman, *Writing at the Margins*, 97).

[24] Kleinman, "Everything that Really Matters," 327.

One can get a sense from this description how central Kleinman has been in the effort to reorient ethnographies to attend to the inherent moral dimension of experience. Yet, even with Kleinman, there is a limit to how well his framework can actually give voice to those experiencing violence and can help us understand exactly what is occurring when a survivor of violence claims they have lost the ability to be "good." Even though he appeals to the term "*moral*" and focuses on experience as inherently moral, he lacks the vocabulary—we might specifically say the second-order vocabulary—to build on this and actually flesh out what such experience is like.

These limits become apparent when Kleinman discusses specific cases. For example, in laying out his idea of experience in his article, "Everything that Really Matters," Kleinman draws on research conducted with his collaborator, Joan Kleinman, on the psychosomatic illness of Chinese individuals who had survived the upheaval of the Cultural Revolution. The Kleinmans draw on their informants' specific bodily complaints to better illustrate to the reader how symptoms and emotions that may seem straightforwardly medical in nature, such as dizziness and exhaustion, can actually be evidence of intersubjective dynamics and be a response to political realities and pressures. His informants, traumatized by the Cultural Revolution, were not permitted by the Communist Party or the state to criticize that period of history, and so they could not express the profound trauma of those times on their lives and the lives of their loved ones. The Kleinmans describe times when some of these individuals came together in small groups, and instead of politically critiquing the state from the basis of their experience, their individual traumas would manifest themselves as a shared, medical trauma. On a broader scale, the Kleinmans argue that suppression of expression turned to physical ailments, which in their aggregate affected society at large, showing how unjust silence could lead to social malaise.

For example, one such informant might be severely depressed by the fact that she could not express what they had experienced and felt, a suppressed suffering that would manifest as *dizziness* and *exhaustion*. As the Cultural Revolution affected the entirety of the country, such feelings had a collective effect and manifested in the body politic, resulting in the "devitalization of social institutions and networks."[25] The Kleinmans argue, then, that such symptoms may reflect medical needs, but more

---

[25] Kleinman, "Everything that Really Matters," 325.

importantly, they also reflect political upheavals, the experience of which such individuals could not talk about easily in the open due to the authoritarian nature of their government.[26]

This is a fascinating read on symptomatology and its connection to political life, yet in a work looking at the importance of the local and the moral—and of the frame, *local moral worlds*—Kleinman does not directly discuss such symptoms as possible indicators of changed moral subjectivity or ability. His theoretical vocabulary, so good at illustrating how different areas of human life and experience are interconnected, as well as the somatic effects of political upheaval, lacks terms that could bring out the moral contours of that experience.[27] He is lacking the more elaborate and elaborated-upon vocabulary found in ethics or philosophy referring to types of morality, moral development, ends, justice, teleology, character, habit, disposition, and goods. Indeed, we have a description of one's experience as dizzying, exhausting, devitalizing, but how did this affect one morally? When the individuals come together within a local moral world, how was the sense of their own morality changed? How did the Cultural Revolution change their sense of goodness and their relationship to it? And, if they changed, what in them changed that resulted in a transformed moral self?

I am tempted to say that these questions are simply not Kleinman's and that putting them critically to his article is not only less than charitable but also a misreading of a researcher's intentions. I resist that temptation, however, because of the stress that Kleinman himself places on the "inherently moral" nature of political experience. Indeed, in the article just discussed, Kleinman describes his informants' experiences as relational in nature, occurring in "a family, a network, a neighborhood, a community."[28] Such relationships are at the heart of what makes one's morality and are the elements that constitute one's *local moral worlds*. Participation in the life of family, of community, is to participate in and eventually internalize

---

[26] Kleinman, "Everything that Really Matters," 325. See also "How Bodies Remember." See also *Deep China*, a work focusing on moral subjectivity and modern transitions, which Kleinman helped edit.

[27] For Kleinman, morality is largely an emotional affair. He talks about "what really matters," and in giving examples, he talks about people who feel better during an illness ("moral regeneration") as well as the uplifting feel of listening to a piece of classical music. It can also be connected with an experience of suffering during disease where change is not welcome but is a harbinger of increased pain (Kleinman, "Everything that Really Matters," 330).

[28] Kleinman, "Everything that Really Matters," 326.

the creation, negotiation, and contestation of values and what matters in life. Yet, the moral dimension of this experience is left unexplored and unarticulated.

My aim in engaging these works in anthropology is to take advantage of the important emphases these researchers place on the local, including the way that the local is conceived as first and foremost a moral ecology, and the ways in which locality interacts dynamically with broader events and institutions involved in conflict. Virtue discourse, I argue, can provide needed vocabulary to bring out even further the moral dimension of what is at stake in violent conflict, while also helping to emphasize the local level of such analysis by providing a way of talking about experience and interiority. Kleinman, for example, is able to bring out the ways in which individual suffering is linked to political upheaval to the extent where suffering is not just a reaction but a strategy for dealing with constrained options for responding to such upheaval. What is lacking, however, is a language to express the dynamics of interiority, which can give voice with more detail to the experience of changing within the context of political violence.

THE NEED FOR *MORAL* SUBJECTIVITY

What form should such a virtue language take? If we return to Maček and Dizdarević's examples, we see people confronted with world-altering violence. The gestures that Dizdarević describes are toxic to the moral basis of social cohesion. Even more, what Dizdarević's lamentation and the anxiety of Maček's informants reveal is a concern with certain horizons of possibility that individuals and even whole communities can aspire toward. Dizdarević called this *civilization,* a world where forgiveness, compassion, and cultivated human interaction (perhaps even care) are the fluid substance of intersubjectivity. Maček's informants refer to the "*normal,*" perhaps closely related to Dizdarević's "civilization," where one can carry on a form of life not only similar to that of the pre-war era, but one also that they could still see on television and radio, one found in neighboring European countries not bound by war. There is longing here, as well as anxiety and anger, for lost possibility. Not only does Dizdarević deny civilization in the face of atrocity, he also denies its possibility. The horizon has set. It seems now an ideal or image that can only indict the unjust and immoral actions of his neighbors. It can no longer call forth one's better nature with its promise. It can no longer serve as a horizon toward which one can aspire. Although there seems to be more hope at least for some of

Maček's informants, there is the fear that such possibilities are fading. The graffiti Maček mentions, "No one here is normal," is provocative and plural in meaning, part warning, part declaration, part witness to a changed world that may not be able to return to what it once was.

There is the collapse here, or the possibility of collapse, of a worldview that one can hope to live into. It is as if old horizons that provided one with the hope of living into certain worlds—worlds where certain ways of being human were made possible—have disappeared, leaving horizons of meager substance and meager consolation in comparison. There is a collapse of vision, not only in terms of perception and the way one sees oneself and others, there is also a collapse of moral imagination, where Dizdarević denies the possibility of imagining certain futures. This leaves an existential orientation, perhaps a subjectivity, that one could call "hopelessness," where hope is the ability to see a way through the present to a different future. It is an orientation, whether it is referred to as moral harm, or moral injury, or some yet-to-be-named matrix of anger, shame, and despair, where such virtues as hope are made much rarer or even impossible to embody.

What is needed to do justice to these accounts is an understanding not just of subjectivity, but also *moral* subjectivity, that reflects conflict between and within persons. It is a subjectivity where such conflicts are seen as significant aspects of moral being and where the moral dimension of the person is emphasized first and foremost in our analyses of violence. This is important, for in trying to understand how situations of extreme political violence can render undone one's felt moral ability, one can run into trouble if their model of the human self takes as normative a stable, peaceful political, social, and cultural space, as does Aristotle and even more recent interpreters of his thought.[29] What is needed, instead, is a way to give an account of such experience that takes seriously the challenges of trying to live into notions of goodness in such social and political turmoil, no matter what those notions are, and so help all of us to understand better the profound, moral stakes involved in violence.

[29] This can be seen in Alasdair MacIntyre's influential work on virtue, *After Virtue*. Although MacIntyre later took pains to show how vulnerable human moral life can be, his vulnerability as understood in *Dependent Rational Animals* is grounded in biological precariousness and not political violence and instability. Indeed, even his original thesis, whereby modernity is typified by moral intelligibility, is not a discussion of the violence of the modern age, but rather of its change in epistemology.

This need can be seen powerfully in the concluding lines of Dizdarević's journal passage. He ends his journal entry with a grim prediction: "Whatever the ultimate fate of several thousand uprooted Sarajevans turns out to be, their deportation has extinguished all gentle and humanitarian feelings in tens of thousands of their fellow citizens. Evil has multiplied in the worst way."[30] The author is talking about a loss not simply in economic productivity, and not even just in lives. He speaks of the loss of the ability to embody moral values—virtues—necessary for social cohesion, for a community to be a community. He describes the dissolution of hope, and in particular the hope of expressing kindness and having that expression returned, kindly, and in kind. It is the loss of a moral world, and in some extreme cases even the loss of the possibility of a moral world. The power of Dizdarević's analysis, then, does not lie in his objectivity or in his ability to stand back from the experience of war. Instead, the power of his writing comes from a profoundly moral analysis of felt experience.

This of course does not discredit other languages and frames of analysis. It does, instead, suggest the importance of reconfiguring such approaches in a way that privileges more local, more personal, more intimate, and even more interior spaces as the ultimate objects of study. It indicates that the predominant social-scientific accountings of violent conflict are necessary but insufficient to capture the stakes that Dizdarević argues are so important to those who go through such violence. Looking at violent conflict in this way, economics and social and political science are important not so much in themselves but inasmuch as they help us get at the dynamics and structures that condition the conflict, which gives rise to the experience of moral and social loss of which Dizdarević speaks. They help to show how disruptions in economics and politics and society affect the lived experience of individuals and communities and what is of most concern to them. Dizdarević's moral language, however, remains indispensable in illustrating what some of the central stakes were for those surviving the political violence through moral terminology, stakes that would have been missed through analyses using different discourses. This would make such social sciences not just human sciences but humane sciences, putting at their heart the experience of deep, transformational suffering and conflict.

---

[30] Dizdarević, *Sarajevo*, 54. For example, both the Croatian foreign ministry and the government of Herceg-Bosna, which was a breakaway region in southern Bosnia-Hercegovina, attempted to deport more than 10,000 Muslim men in 1992. Within one summer, 45,000 to 50,000 Muslims had been "cleansed" from the Mostar area alone (Vulliamy, *Seasons in Hell*, 323).

Three more points should be noted before proceeding. The first is that I refer in these pages to specific contexts of violence, particularly political violence, war, genocide, prolonged siege, and so on. These are more overt and extreme contexts in which one's life possibilities are significantly curtailed. *Violence* can refer to a wide variety of actions and experiences, from structural violence, which can have a severe impact on one's life yet also not have a particular agent who carries out the violence, to battery and physical abuse. Such forms are connected and interrelate. That I do not spend much time on different forms of violence does not indicate a stance that the approach developed in this book cannot be used in other ways, or that further work needs to be done. Instead, it is a beginning where we probe some different ways that virtue can help in understanding the complex dynamics of violence.

Second, notions of virtue and virtue ethics more specifically have been at the center of various debates within philosophy and theology since the middle of the twentieth century. First, the field of virtue ethics and the use of virtue discourse have, for the most part, been concerned with questions within the discipline of philosophy, including meta-ethical questions about the concepts and methods of ethics itself.[31] In the last half-century, it has been drawn upon in Christian, mostly Catholic, theology, as an alternative to philosophies focused on means, duty, utility, or outcomes. These projects are not just normative but also prescriptive; they describe in order to propose and even to persuade. I make no arguments for or against virtue approaches to ethics, nor do I propose a prescriptive ethic myself. Instead, virtue discourse is here engaged as a way to give a norm-laden voice to particular experiences, their conditions, formation, and genesis.

This work, then, is not an ethic. Indeed, the language of the virtues is used in a significantly different way in this study than in standard philosophical arguments. *Ethic* implies action, prescribing certain approaches, which is not intended in my use. *Discourse*, on the other hand, implies a

---

[31] There are, of course, exceptions. Again, Alasdair MacIntyre engages in virtue ethics to critique the project of the Enlightenment, liberal rationalism, and capitalism, proposing a new approach based on what he sees as a neo-Aristotelian-Thomist tradition. Lisa Tessman and Christine Swanton have been among the first to try to expand the application of virtue beyond its more recent concerns. There have also been other virtue ethicists trying to explore its application in other fields, including those in the edited volume, *Working Virtue* (Walker and Ivanhoe, *Working Virtue*). This is a further, helpful extension of virtue ethics, though unlike the present study, they do not attempt to use virtue discourse to help analyze experience and the social repercussions of violence and social change more generally.

shared way of speaking of certain issues among a certain group with shared norms. The present work is concerned with finding better ways of talking about the experience of violence and how it affects the individual's moral architecture and sense of self. To do so, I make use of what we could refer to as virtue discourse or language. Although always normative, virtue is used not only in ethics and philosophy but is also colloquial, used in society at large, as well as specific communities, such as theological communities and with regard to religious practices. *Virtue discourse* speaks, then, to a wider category than *virtue ethics* indicates.[32]

The key assumption that undergirds this distinction is that those engaged in ethics, particularly certain virtue ethicists, in order to make certain claims about the nature of humans and their development within society, employ languages that they inherit, change, and engage that enable them to speak about these issues with a level of sophistication concerning the moral dimension of human life not found in other discourses. These languages have a strong descriptive element that is used to make such claims, an element that we should be able to use for multiple purposes, not just those of a prescriptive ethic. Indeed, we should be able to apply these languages and their understanding of human individuals and communities without needing to adopt ethical agendas and goals. They can help us to better understand what people have experienced as moral subjects. Virtue discourse is most commonly used in scholarly discussions, and even in more colloquial, everyday uses, to argue for how the world is, how people should be, all based on an account of the world and human nature. In other words, it is the articulation of what anthropologists like Clifford Geertz call a *worldview* and *ethos*, which includes assumptions of

---

[32] *Virtue ethics* itself is a contested category. Martha Nussbaum, who was part of the spread of virtue ethics in the late 1970s through the end of the 1980s, eventually questioned the salience of such a concept, arguing that it makes little sense to define a category of ethics called *virtue ethics* when thinkers representing other categories, such as Kantians and Utilitarians, also use virtue. Nussbaum argues, instead, that we categorize virtue ethicists according to what they are arguing against (such as "anti-Kantians") or the philosopher they draw on ("Neo-Humeans" or "Neo-Aristotelians") (Nussbaum, "Virtue Ethics: A Misleading Category?" 200–1). There is much to be said about this, yet labeling one's thought as "anti-" privileges the critique over the constructive element, and the modifier "Neo-" is not ideal as it can reduce one's thought too quickly to that of another, opening one up to dismissal. The fact that many virtue ethicists see a closer relationship to each other's projects than that of others should also be taken into account, which Nussbaum does not do. In this way, I acknowledge the problems with the category yet also affirm that there is a substantial and fruitful discourse that the umbrella term "*virtue ethics*" allows.

how we and the world are, as well as a connected evaluation of how we should be in the world, and even at times how the world or society should be.[33]

This is a key aspect that separates my approach from that of theologian Aristotle Papanikolaou. Papanikolaou's work is more straightforwardly theological and Trinitarian in his discussion of virtue and violence, but he also engages in analytical virtue ethics, or, as he puts it—drawing from Adams—"the ethic of virtues."[34] Despite my respect for his project, I hesitate to follow him. Part of this has to do with the readiness, even eagerness, of virtue-ethical traditions to name the virtues and the vices, as if they are received and given. This is not necessarily something that Papanikolaou does, yet Lisa Tessman's critical virtue ethics, in which she draws on womanist and feminist writings to reclaim anger as a virtue, although not an unproblematic one, unsteadies the notion of the virtue and vices that we have received. Indeed, ethicists such as Katie Geneva Cannon, and thinkers such as bell hooks, have put forward as virtues qualities that have historically been vices. And they have done so by putting forward the perspective of historically marginalized voices and communities that even at this late date have simply not been treated as central to the reigning virtue ethical traditions in theology and philosophy. Indeed, as a queer man myself, I have been subjected throughout my life to a virtue ethics that has been weaponized against my very being and my mode of loving. All of this gives me pause before deploying the virtues in any analysis.

Investigating such issues is an important project and one that needs its own space beyond these pages. Whether virtue ethics, and ethics formed with a focus on the virtues, can be reworked for discussing moral injury and the way in which violence effects subjectivity, however, is a question that, I argue, is asked too soon. Instead, the argument in these pages is that the language of the virtues need not be beholden to formal virtue ethics and virtue-ethical traditions. They can be creatively adopted and reworked to articulate the moral dimension of experience. This does not deny interesting projects that are well-grounded in scholarship and theology, such as Papanikolaou's. It marks, however, an important distinction between a project investigating the creative adoption of virtue discourse to analyze the moral dimension of experience—one that does not intend on

---

[33] Geertz, *The Interpretation of Cultures*, 87–141.

[34] Papanikolaou, "Trinity, Violence, and Virtue." See also Papanikolaou, "Learning How to Love." My thanks to Papanikolaou for thinking through this distinction.

creating an ethic from such activity—and seeing such experience through the lens of an established ethic that works within a quite old tradition of theological virtue.

The virtue ethicist Rosalind Hursthouse has written that "… in evil times, life for most people is, or threatens to be, nasty, brutish, and short and *eudaimonia* is something that will be impossible until better times."[35] This is also what some, perhaps many, who survive extreme political violence also claim, though not always with the resources to articulate this truth so readily. Such survivors even exceed Hursthouse at times, making the claim that violence can undo the very possibility for what we might think of as a good, meaning flourishing, life. Indeed, one's personality does not stay constant until those "better times." People change under brutish regimes and the threat of violence. If one flourishes, then, it will not be as the same person but as one who has changed through violent experience. And, for some, it may feel like *eudaimonia* is no longer possible, or something they are no longer worthy of. This work is an attempt to help us talk about these phenomena better, and, in this respect, is a beginning.

The present study, then, uses virtue discourse, specifically the discourse and imagery that Iris Murdoch formulates, to create an understanding of moral subjectivity, a virtue hermeneutic, so that we can speak of what happens to someone when they have survived political violence only to feel they can never be good again. This will constitute an interpretive frame that can be deployed and can empower us to speak of this in a way that forefronts the moral dimension of experience and the self as a moral subject embedded in a larger world that conditions moral subjectivity to a large extent. A virtue hermeneutic provides us with concepts, a vocabulary, but most importantly a model of the self that privileges experience, particularly its moral aspects.

What this frame interprets particularly is *moral subjectivity*, providing the concrete language and concepts to describe the experience of being a moral subject that I have argued is lacking in current work on violence and subjectivity. Moral subjectivity, as we can tentatively define it, is the experience of being a moral subject, one who is constantly orienting the self toward and away from culturally constructed and internalized images of the human person, working toward certain goods, and negotiating between various locations of account, obligation, and responsibility. The

---

[35] Hursthouse, *On Virtue Ethics*, 177.

formation of the self is understood as aspiring toward what is good, however that is understood. What is at stake are the ways in which such aspirations can be sometimes strengthened, sometimes challenged, and often reinterpreted, reconstituted, and sometimes even overthrown.[36] Moral subjectivity, then, is an interpretive frame that will ground our inquiries in such a way that the moral dimension of experience, particularly the experience of political violence, is paramount. Such subjectivity includes evaluation and concern. It is not, however, simply the categorizing of the world into good and bad, right and wrong, the straightforward creation and application of value-oriented rules, nor particular instances of judgment, deliberation, and choice. It is meant to capture, instead, a more holistic sense of selfhood as life lived among others defined primarily in terms of moral development and transformation. This is a subject, however, in existential motion, in which moral development is a central concept for understanding such subjectivity. Every moment, action, thought, and decision accrues to such development and alters one's subjectivity, just as changes in one's environment and in one's networks also alter that subjectivity.

Finally, this work relies heavily on the experience of those in a particular war zone who have experienced genocide directly or indirectly. The authority of their experience, then, is recognized, yet also I recognize the fact that personal experience requires interpretation and is not a uniform thing. As we have already seen, one's subjective immersion in a situation, and reflection on it, may prompt both more questions than answers, as well as a call for assistance in naming and understanding what one has seen, heard, sensed, and touched. This means that although such experiences are privileged in these pages, one would miss the point to see the articulations of witnesses and informants as having complete understanding and mastery of their experience and reactions. Humans are not transparent to themselves, and so we are not automatically authoritative about our own motives or even our deep feelings. Knowledge of self is created and mediated, and it often requires engagement with others.

Particularly in the experience of political violence, there are often many important reasons for ignorance of one's own experience. Arthur and Joan Kleinman, for example, have argued this in their work in post-Cultural Revolution China. Trauma and continuing political violence can threaten

---

[36] This understanding is derived both from Charles Taylor and Iris Murdoch, particularly in *Sources of the Self* and *The Sovereignty of Good*. As we delve more into Murdoch's philosophy in later chapters, we will build on this notion of the moral.

those who speak, making survivors articulate their histories of violence, either consciously or unconsciously, in terms of trauma, illness, and other symptoms, and when they express their experience verbally, they may not have a vocabulary to fully represent the depth of the moral stakes involved. Ivana Maček also describes how her informants tried to make sense of the war, and yet how just as often they would fail in their attempts. Another of Maček's passages further reflects the difficulty of those who experienced the violence to give an account of that very experience:

> People who could easily give me a sophisticated political analysis one day would the very next day express bewilderment and ask me to explain to them why all this was happening. Something that had made sense could suddenly become meaningless; what had momentarily seemed normal could crumble into nothingness.[37]

This is important, as Maček is calling attention to a central aspect of the loss of goodness and moral ability, to which her informants and others going through violence have testified. Something about this experience is deteriorative to meaning, to norms, to the comprehensibility of one's life, and seems to exceed one's common vocabulary for grasping it. Important to note, too, is the way that self-understanding and apparent authority over the meaning of one's experience can change with time and circumstance, making the process of meaning-making and knowledge creation that much more difficult and dynamic.

We may well, and may often, struggle to find a vocabulary to express new experiences, particularly ones that feel rupturing and that alienate us from our own world, as in political violence. And, in this process, we are not only interpreting our experience to others but also making sense of it to ourselves, engaging in a process of self-knowledge.

IRIS MURDOCH'S MORAL PHILOSOPHY

To help develop such an approach, I turn, then, to another resource, the moral philosophy and philosophical anthropology of Iris Murdoch. Murdoch is an interesting figure, better known for her literary career than for her philosophical contributions, even though her first writings were on

---

[37] Maček, *Sarajevo Under Siege*, 9.

philosophy.[38] Murdoch was influential for many key philosophers and ethicists of the latter half of the twentieth century, including John McDowell, Martha Nussbaum, Stanley Hauerwas (in theological ethics), and Alasdair MacIntyre, perhaps the best-known virtue ethicist of this period. She was one of a group of women, along with Elizabeth Anscombe, Philippa Foot, and others, who helped the mid-century resurgence of virtue in philosophy.

Her influence on its own makes her a compelling focus, yet Murdoch's thought is particularly helpful in developing an understanding of moral subjectivity made "thick" through the rich, particular, prescriptive vocabulary found in virtue theories, what I call a "virtue hermeneutic." Murdoch was not a Christian or religious ethicist, but was what could be called "post-Christian." Murdoch drew from religious ethical discourse, reinterpreting such concepts as "grace" and "transcendence" for a post-Christian, secular culture.[39] That the eclipse of religion has not come to be is not important here. Instead, Murdoch aimed in part to create a vision of the moral life grounded in a metaphysics that allowed certain resources and insights of theology to be reinterpreted and applied within modern, secular philosophy. This makes Murdoch a bridge thinker, joining the resources of Christian ethics and theologies with modern philosophy. Indeed, much of her final opus, *Metaphysics as a Guide to Morals,* is clearing room in philosophy for a non-theistic virtue approach that nevertheless borrows from a theistic, salvific, theological tradition. In the modern vernacular, she is bridging the "religious" to the broader, less confessional "spiritual." Whether one thinks that her metaphysics works in a broader sense, the language Murdoch develops to speak about this metaphysics and how individual characters are formed is a rich discourse with the potential to reframe and rearticulate experiences in new ways that reveal the moral dimension of such experiences.

In addition, Murdoch will be helpful through her understanding of the human person, drawing as it does from virtue language. This understanding sees the moral life involving, first and foremost, notions of desire, hope, vision, and imagination, as well as the very real possibility of failure and despair, which give us an almost visceral entrée into human interiority, consciousness, and felt experience. For example, in referencing Dizdarević and Maček's ethnographic work, I have been describing people for whom there

---

[38] Murdoch was one of the first to introduce Jean-Paul Sartre's philosophy to an English-speaking audience. See Murdoch's *Sartre: Romantic Rationalist.*

[39] Murdoch, as we will see later, was also influenced by Zen Buddhism.

has been a collapse in their perceived ability to experience themselves as good or possibly being good. This is a focus on the individual as one who is malleable to changes in the larger society, one for whom the moral dimension of living is most important for understanding the harm and loss of ability they have experienced. I am talking about, then, a primarily *moral* self for whom tension and development are central characteristics. In Murdoch's understanding of the self as a "field of tension," the individual is pulled by different fields, communities, institutions, loyalties, and spheres of value, all with different obligations, often incommensurate to or in competition with, and all providing different yet worthwhile, even necessary, goods. Being in competition, this field and its different modalities—*axiom* (political), *Eros* (personal desire), *duty* (beyond virtue), and *void* (experiences that challenge moral intelligibility)—as well as the goods they supply, necessitate a constant evaluation and negotiation that makes up the moral life and the activities that constitute our moral development.

This idea of moral subjectivity defined by tension will be central to this work's argument. Murdoch argues for a subject embodying tensions among competing loyalties, incommensurate goods, and differing identities. The change and tension at the heart of being a moral subject, then, not only propels one's development as a moral subject but is also the condition and ground of one's tension snapping or slacking, enabling a form of despair or feeling of moral loss, in which one's very identity can be at stake.[40] Her fourfold representation of subjectivity—of *axiom, Eros, duty*, and *void*—enables us to chart and track changes in moral experience. Within this representation, the most important modality for my inquiry, and the one that I will mostly use, is Murdoch's concept of *void*. In certain extreme situations of violence, it can become impossible to honor adequately one's responsibilities and loyalties, and to attend to all and everything one loves and cares for. The value-filled world becomes endangered, yet one does not have the resources to respond with care to all "areas of account," as ethicist H. Richard Niebuhr would put it.[41] This is critical, as the inability to respond equally and with care to these different modalities can lead one to despair. The impossibility of living into these competing

---

[40] I have discussed the connection between morality and identity but not fully enough. I will do this further in the following chapters. Philosopher Charles Taylor has already asserted the relationship by stating that our identity is informed by whatever our understanding of good is (Taylor, *Sources of the Self*, 27).

[41] Niebuhr, *The Responsible Self*, 63.

horizons may lead one to feel that morality as a project of constant development and constant mediation between competing goods, loyalties, and horizons is no longer possible.[42] The task may not be worth it, or the subject may not be capable of the work. The tension can grow limp or snap, leaving only the pull of void, which then draws one's moral subjectivity toward despair and meaninglessness. If one's moral subjectivity is dominated by the modality of void for long enough or deeply enough, this can result in moral injury, a long-term subjectivity in which one feels they are no longer able to strive for or access visions of goodness.[43] One can feel that a new inability to be "good" again or that, perhaps, talk of the "good" has always been just a sham, after all. Such loss can, indeed, be experienced as an inability to orient oneself to the good, where such a teleological capacity is felt to be weakened or diminished beyond the hope of repair.

In spite of the large amount of scholarship on Murdoch, her understanding of the self as a field of tension has received relatively little attention. Even *void* is dealt with as more of an aside or engaged to understand some other aspect of Murdoch's thought. Those who do look at void—such as the ethicist Maria Antonaccio and philosophers Stephen Mulhall and David Robjant—do so to examine Murdoch's notion of the *Good*, goodness, and more general interpretive questions.[44] What I propose to do with *void* is to apply it as a way to interpret specific experiences of extreme violence.[45] The main goal, however, will be to draw on Murdoch's virtue-grounded metaphysics, particularly from the final chapters of

---

[42] Some of the terms that I use to describe Murdoch's thought, such as *horizons,* I take from philosopher Charles Taylor, a student of Murdoch. Taylor, who shares Murdoch's universal conception of *goodness,* along with a sense of its content as broad and culturally relative, describes moral *horizons* as the moral ideals toward which individuals strive. And Taylor's notion of horizons can help further underscore what is central for Murdoch in her framing of the self as a field of tension: the importance of striving toward the horizons embodied in the different modalities that make up that tensile field. I will sometimes refer to Taylor to bring out aspects of Murdoch's thought, but articulated in ways that may be more helpful in adapting Murdoch's thought to a more engaged, analytical application (Taylor, *Sources of the Self,* 17, 27–28).

[43] I take the notion of virtue as a form of striving from Julia Annas's virtue ethics (Annas, *Intelligent Virtue,* 18, 52).

[44] Antonaccio, *A Philosophy to Live By,* 43, 182–184; Antonaccio, "A Response to Nora Hämäläinen and David Robjant;" Mulhall, "All the World Must Be Religious;" Robjant, "How Miserable We Are, How Wicked."

[45] This move to application in Murdoch's thought is something that Maria Antonaccio has called for in her *A Philosophy to Live By.*

Metaphysics as a Guide to Morals, to develop a critical interpretive frame drawing on virtue discourse—a virtue hermeneutic—to be used to account for and articulate the moral dimension of experiences of violence and subjectivity.

Murdoch, then, provides a range of concepts that can help articulate and account for moral injuries of the type Dizdarević and Maček describe. These concepts focus on an understanding of the self as a moral self and help create a representation of the dynamics of life as a moral being. In particular, I look at Murdoch's understanding of the self as a "field of tension," understood as a self for whom life's moments are primarily moral in character; her emphasis on interiority that is nevertheless open to the social; her view of ethics as involving perception and vision and the importance of horizons and striving in morality; and her understanding of void as a phenomenological concept. These concerns provide her with an array of tools that can help in better understanding the suffering of political violence. Murdoch's understanding of the human as a tensile moral subject, which I will examine through the course of the book, pulls together in one representation the nature of the self as moral subject, and will form the basis of this study's representation of moral subjectivity, which we can use to better articulate and account for changes in moral architecture and development through violent experience.

These elements come together in a frame that one can use to analyze experience and provide an account of such experience. This vocabulary and model can enable both survivors and scholars to articulate the experience of overwhelming violence. And it can create the foundation for arguments that can make such experience more central to discussions of political violence and its aftermath. These forms of phenomenological harm reflect changed societies and altered subjectivities, the very locations in which post-conflict work is conducted. Without accounting for such experience, decisions and policies affecting whole populations cannot reflect adequately the needs and challenges on the ground. Murdoch provides us with the capacity to talk about subjectivity unmade.

I supplement Murdoch's representation of the moral subject with some of the anthropological concepts gleaned from Kleinman and others—such as the *everyday* and *local moral worlds*. These concepts will help us locate more precisely the social and cultural position of the moral subject experiencing violence, opening Murdoch's resources, which are more individually and phenomenologically focused, to consider the ways in which social structure, institutions, and the interactions between various levels of social identity, action, and value condition the moral life and subjectivity. By connecting Murdoch with work in the social sciences, my aim is to create a ready available approach for social scientists and to demonstrate how such a metaphysics and moral philosophy can be fruitfully used to understand human action, motivation, and experience in general. We can then use this representation of the moral subject to see how and in what way individuals come to feel they have lost their ability to orient themselves toward the *good* by using Murdoch's modalities and conception of the moral self to analyze accounts of moral harm.

MAP OF THE PRESENT WORK

The following chapters explore a particular way to articulate and account for the experience of a moral subject in the midst of violence, and how violence can deteriorate one's felt sense of moral ability and orientation toward the good. Throughout I will continue to refer to the case of the Bosnian War to provide examples and cases. The study starts with a broader scope and then narrows as it progresses through each chapter.

I begin by looking at subjectivity, a much-used but rarely defined term. Chapter 2 explores an understanding of subjectivity with the potential to capture the experience of political violence survivors. Drawing on anthropologist Sherry Ortner's definition of subjectivity, I present an understanding of subjectivity that instead of emphasizing will or reason emphasizes emotion, even imagination, as well as the influence that culture and institutions have on one's selfhood. Ortner's understanding of subjectivity will provide a foundation, yet I will argue that her conception is not sufficient for the purposes of this project. As Sayer has argued, and as I have illustrated through the work of Kleinman and Maček, the moral dimension of experience can too easily be elided without an already posited methodological emphasis on the moral.

For this reason, I will argue for a frame that is a *moral* subjectivity, one that embodies an understanding of experience as first and foremost moral. This understanding of moral subjectivity, in addition to other characteristics, will make moral development a central characteristic of being a moral subject.

Even so, moral subjectivity understood in this way is still limited in its descriptive ability. As moral subjectivity is a thin, abstract notion, it requires rich, descriptive images of the human, as well as a vocabulary, to make it thick and applicable to real life situations. Beginning with the third chapter,

I develop a virtue hermeneutic that fills in such details. Chapter 3 specifically examines Murdoch's philosophical anthropology, her account of moral development, and her metaphysics and how these three categories are related. I do this to better understand the structure of moral subjectivity in order to illustrate how that structure can leave us vulnerable to sudden political violence and change. Specifically, I illustrate how one can feel a sense of responsibility that comes from such vulnerability, even if objectively the individual had no power over the situation at hand. This can help explain why moral persons can feel moral responsibility, and eventually moral inability, even when acting ethically in difficult situations.

Chapter 4 focuses on the self as a field of moral tension, or what I call "the tensile moral self." The different modalities of ethical being that Murdoch describes—*axiom, duty, Eros*, and, *void*—provide a representation of moral experience and the way individuals create value. These different modalities, each arguing for different ideal images of the human, are not always compatible, and one can feel failure when one modality is attended to at the expense of another. During times of violence, this can become exacerbated until, experiencing too many situations with no good choices, one may come to feel either that moral effort is useless or that one lacks an ability to be good in such situations. Murdoch's description of moral development and change is so important because she provides words and images to describe the *experience* of such change and development, an experience of one's interiority. Combining this with Kleinman's understanding of *local moral worlds*, I demonstrate how this combination can articulate and account for the dynamic context in which moral life is lived, as well as the subjective experience of such life.

Chapter 5 looks specifically at void, and I argue for an understanding of void that reflects extreme political violence, and so can help us better understand the experience of moral life that survivors attest to. Void here is an experience in which morality becomes unintelligible and where one no longer feels able or worthy of striving toward images of goodness. Using examples from the Bosnian War, this chapter examines the ways violence so affects social life that one is no longer able to be the same person and no longer able to engage in practices that allow them to strive for goodness. It is a description of how a person's moral subjectivity can be undermined, how it can become dominated by void, in a way that damages their perceived moral ability and very identity.

This leads to Chap. 6, which engages more fully with the idea of moral injury. Having used this term earlier in the book, I argue that the harms

Maček and Dizdarević describe are moral injuries, understood as harm done to one's moral subjectivity and ability to imagine oneself as a morally capable person. The conception of moral injury so far developed is compared to what I argue are the three main discourses of moral injury present in philosophical and psychological research: clinical (mostly psychology), juridical (mostly found in philosophy), and structural (mostly liberationist and feminist). The chapter ends by locating this work's understanding of moral injury among these other discourses and creating, at the end, an understanding of moral injury understood through the frame of moral subjectivity and virtue discussed thus far. The analysis ends with a more robust understanding of the morally injurious subjectivities witnessed by Dizdarević and Maček.

The conclusion returns to the original argument of the text, namely, that ethics can contribute important insights and theoretical resources to discussions, such as war and political violence that are often seen as falling within the purview of the social sciences. The book is an argument not only about applying ethics in a different mode but also about methodology and how to approach certain experiences and events in ways that do not ignore their moral dimension, a dimension important to survivors themselves. The chapter and the book end with a reflection on the consequences of this work for methodology and for humanistic approaches to practical issues, making a final argument for the broader relevance of the ethical humanities and philosophy.

CHAPTER 2

From Subjectivity to Moral Subjectivity

As philosopher Alasdair MacIntyre has commented, where one begins an inquiry has a great effect on the conclusion of such an inquiry. This insight applies here, as the success of articulating and accounting for the experience of violence and its influence on the individual is dependent on how the *self* is understood. Specifically, we need to form an understanding of the self that can reflect the complexity of experience of those such as Dizdarević's and Maček's informants. To do so, I turn to the discourse of *subject* and *subjectivity* to describe the human individual and her experience. Subjectivity includes in its definition an understanding of the self, if you will, as highly conditioned by history, culture, politics, and social structure.[1] This is important because the structural and social context

---

[1] There are many ways of framing an individual's experience of perception of their own subjecthood and the world. Words that reflect an individual's point of view and history include terms such as *self*, *agency*, *being*, and even *consciousness*. Each term has its own histories and assumptions and each reflects attempts across centuries to capture aspects of existence and the experience of being emotional, aware, and animate. I could employ any one of these terms in the present work, yet my concern with each is that, if overemphasized, they lead away from the moral experiences just discussed. *Agency*, for example, can overemphasize one's will and ability to act as an individual. Although agency is an important part of wartime experience, the feeling that one lacks such ability is also an important aspect of war. This is particularly true in a study such as this investigating the feeling that one is radically unable to act in certain virtuous ways. *Self* reflects more psychological understandings of the person, as does *consciousness*, with its strong emphasis on interiority. *Being* has a long history, with much baggage, and each term—*being, self, consciousness, agency*—tends to assume an image of the individual as something of an essence, whether this essence is understood through nature,

© The Author(s) 2019
J. Wiinikka-Lydon, *Moral Injury and the Promise of Virtue*,
https://doi.org/10.1007/978-3-030-32934-1_2

inherent in subjectivity's definition and history make it useful in reflecting the dynamics of violence on an individual. Any such study requires an understanding of the human individual as highly conditioned socially, as we are discussing how violent changes in those conditions transform the subject. And this is an understanding that the frame of *subject* and *subjectivity* can provide.

SUBJECTIVITY AND AGENCY

What specifically, then, do I mean by subjectivity in the context of this inquiry? What elements are critical to building an interpretive framework, a hermeneutic, that can adequately describe and account for the types of harms and loss that Dizdarević and Maček express and document? Subjectivity is a difficult word to define, and different writers use it in different ways for different purposes.[2] What is important for my purposes here is to balance within subjectivity the idea of people as having agency and of being conditioned by norms and social forces, structures, and institutions.

The concept of agency, for example, de-emphasizes the conditionality of selfhood and affirms, when employed within a theory of subjectivity, how individuals regularly experience themselves as driving instead of being driven. It is an important term because it reflects the fact that we *feel* like moral subjects, not just objects that are acted upon. Beyond philosophical discussions about free will and determinism, agency reflects the claim that most people feel agentive in their everyday lives, or, at the very least, that it is an aspect of being a living being. The concept of agency as a central dimension of what it is to be human balances the experience of being a moral subject with structures and events that seem more deterministic in relation to the individual. At the same time, individuals are not autonomous, and although individuals may feel agentive, one's broader context throughout life shapes what is possible to a person in terms of action,

will, perspective, or even soul. This tendency can muddy attempts to talk about transformation and change.

[2] Philosopher Amélie Oksenberg Rorty argues that there are at least six different modalities in talking about subject: "(1) first-person, (2) individuated, (3) self-referential, (4) authoritative veridical report (or expression) of an (5) occurrent (6) mental state (sensation, emotion, thought)" (Oksenberg Rorty, "The Vanishing Subject," 44).

thought, and imagination. This is important, as this social-structural emphasis within some understandings of subjectivity pushes back against the Western cultural narrative of personal independence and possible, extreme understandings of the individual as atomistic or unbounded socially. It is an understanding of agency that does not insist that individuals have the power of will and action at all times. Instead, it can also be understood to illustrate when one feels they do not have agency or have lost the ability to act or think or imagine in ways necessary or important to an individual. In other words, the idea of agency is just as important in illustrating the feeling of lacking or having lost assertion as it is in giving contour to the experience of asserting one's identity and plans within the world.

I take this understanding of agency from anthropologist Sherry Ortner's discussion of subjectivity. For Ortner, agency is a part of subjectivity, with subjectivity providing the basis of agency. This framing underscores what Ortner means by agency, specifically, that it is not a reference to some rational "originary will" traditionally understood. That is, the will is not the seat of personhood, decision, and action as is sometimes claimed. It is not an independent ground of ethical being unencumbered by histories and nomothetic structures such as race and gender, and outside forces. Instead, agency refers to an understanding of action coming from "desires and intentions" and arising from "within a matrix of subjectivity."[3] This allows a conception of action and agency that is neither fully determined nor self-originary, one that arises as an experiential aspect of being a subject in the world.

Further, this tension and confluence of agency and subjectivity is an innately political one. Ortner's understanding of the self is a "subject caught up in a world of violence, state authority, and pain, the subject's distress under the authority of another."[4] It reflects that history of subjectivity already mentioned, one formed in negotiations and conflicts over power. We can include within this the understanding of the political issues

---

[3] Ortner, "Subjectivity and Cultural Critique," 34. Her choice of "desires and intentions" is reminiscent of Geertz's "moods and motivations."

[4] This is important to my goals in this work, although the emphasis on the political does not start with Ortner. As Tanya Luhrmann writes in her recent work on subjectivity, anthropologists, when referring to subjectivity, usually assume a political subject, particularly one that is caught in violent or oppressive environments. See Luhrmann, "Subjectivity," 345.

of political violence, the collapse of state authority, the contestation of authority, and the strategies employed to do so, all of which makes Ortner's understanding of subjectivity relevant to the present study.

We can see from Ortner, then, and from the conversation above, that subjectivity is highly complex. This is welcome, however, as it reflects the complexity of individual experience. What happens to an individual, whether through extraordinary violence or everyday events, cannot be understood without understanding the way that being an individual is highly conditioned. Without a sense of agency, however, the emphasis on conditionality can become a simplistic, inaccurate determinism. Both are needed.

However, these different aspects of being a subject area are not easily identified or understood in practice. An individual can feel responsibility for (having agency over) an event or outcome, while a less involved party might observe the situation and argue that in fact the individual really has no power to influence events. Likewise, one could throw their hands up and claim impotence, when it might be plain to all of those around her that she was the only one with power to change events. This gets to an important distinction in this discussion: the experience of being a subject is a different viewpoint from the frame of an observer, analyst, or researcher. Thus, when I claim that being a subject entails both agency and conditionality, this does not mean that one experiences her circumstances in a way reflective of Ortner's schema and mine. Instead, such schemas can help both a researcher studying violence and the individual experiencing violence to understand better the ways in which events and social upheaval have affected the individual. It should help survivors put words and concepts in dialog with their feelings, impressions, reflections, and so on to create an account of their experience that is not necessarily less complex simply because an account has been given. Instead, it may bring into greater relief how complex the constituents of one's experience truly are, and, in so doing, put one's own agency into better perspective.

Subjectivity is a way, then, to begin approaching another's experience, and, with the tools and concepts that we will later develop from Murdoch, to provide an account of that experience. And any such account should be revealing, for the observer and the individual whose experience it is.

SUBJECTIVITY, EXPERIENCE, AND INTERIORITY

We also need to see subjectivity, in addition to its agentive dimension, as a way of describing experience more generally, as well as the conditions that give rise to experience.[5] More than just accounting for the dimensions of agency and condition, subjectivity is meant to capture the sense of being a center of thought, sensation, feeling, emotion, and value, among others. It reflects the impressions of a being that interacts with the world and organized human life, both influencing and being influenced by that world.

I have already touched on what *experience* could mean—a much-used but elusive category. It might seem so basic to subjectivity and being human that it is very difficult, perhaps not possible, to define it by appealing to more basic concepts. One way to get at this concept, however, is to follow Kleinman and see experience, first of all, as a matter of social level. Experience is an aspect of local moral worlds, meaning that one surely can experience larger world events, but one does so through one's local existence and through local structures, institutions, and norms. We can, however, say more to theorize experience and better elucidate the phenomenologically transformative power of political violence.

Kleinman, for example, describes experience "as the intersubjective medium of social interactions in local moral worlds. It is the outcome of cultural categories and social structures interacting with the psychophysiological process such that a mediating world is constituted. Experience is the felt flow of that intersubjective medium."[6] There is much in this definition that is helpful, and I sympathize with Kleinman's use of the word "medium" to describe experience. Experience is, again, such a basic concept that it is difficult to define it in terms of other concepts. "Medium," however, may give the wrong impression that experience is a thing, a substance, or something like ether, through which the human moves. Instead, experience is the interaction between what we think of as the subjectively individual human in a pluralist social landscape and how that interaction affects the first-person embodied perspective, as well as how it affects the basis of and what conditions such a perspective. In other words, experience can be understood as the subjective, embodied (felt) interaction with and between self and world, mediated through already existing memories, histories, cultural narratives, and social structures that create an always-transforming and transformative worldview and ethos.

---

[5] Mansfield, *Subjectivity*, 6.
[6] Kleinman, *Writing at the Margins*, 97.

Of course, this definition uses terms, such as "self" and "world," that themselves are contested, but the point I want to underline is a practical one. As the focus of this study is to better understand how political violence can transform one's understanding of who one is (identity), and what one can do and strive to be (moral ability), any such theoretical discussion of experience should be deemed adequate insofar as it helps enlighten practically the transformative outcome of violence that those such as Dizdarević have described. To this extent, I argue that it is through what we call experience that persons are transformed. And such experiences are not so much a medium as they are an aspect of intersubjectivity. In particular, to speak of experience is to see the way in which intersubjectivity affects the individual subjects, their perception, self-understanding, and all related "psychophysiological processes," all within the context of local moral worlds. Experience is, then, to be a bit circular, the first-person experience of being a subject in intersubjectively structured local moral worlds and how existing as a part of these worlds transforms one's subjectivity, one's understanding of one's self, one's identity, and one's conception of one's own abilities.

There is a danger in such abstract definitions, however, of losing sight of the forest for the trees. It is critical to emphasize context and the conditionality of human existence. Dizdarević's and Maček's informants, after all, speak of visceral, highly emotional memories. To help understand how those memories are even possible, we need to have an account of the way in which subjectivity is always intersubjective and how the individual is always social. At the same time, the call to try and address such issues comes not from such a theoretical standpoint but rather from the experience of those in pain and who are claiming to have lost something central to their lives, a loss significant enough that it should make everyone take pause. Ortner, for example, emphasizes the conditioned and structural aspects of experience and does so by appealing to practice theory and theorists such as Pierre Bourdieu. She also, however, emphasizes the importance of emotion and what could be called "interiority." Intersubjectivity needs to be affirmed, but her understanding of subjectivity is always "more" than that.[7] It includes some understanding of the inner world that cannot be reduced simply to an extension of some externalized reality. Ortner resists a simplistic behaviorism that would claim anything worth knowing about experience is transparent to

[7] Ortner, "Subjectivity and Cultural Critique," 34.

external observation and that there really is no inner world to speak of. This internal dimension is not a secret, essential aspect of being that is an unimpeachable authority of one's experience, making the subject's statements concerning herself impervious to engagement or critique.[8] Rather, Ortner is insisting that the complexity and nuance of human experience exceeds simplistic notions of *inner* or *outer*.

Anthropologist Tanya Luhrmann, in particular, emphasizes the emotional aspect of Ortner's subjectivity, and argues that subjectivity should largely be understood as the *emotional life* of the subject.[9] Luhrmann picks up on Ortner's reference to Raymond Williams' definition of subjectivity as "structures of feeling" to argue that Ortner's take on subjectivity is largely synonymous with this emotional life. She does this to argue more generally for the salience of psychology as a way to create a clearer definition of subjectivity for anthropologists, a project that differs from mine here.[10] Yet, this understanding of subjectivity as necessarily grounded in emotions creates an opening not just to a deeper understanding of subjectivity but also to the experience of *being* a subject. Bringing in emotion and sensation adds another level, as well as a vocabulary, to help one articulate what it feels like to be a subject at the nexus of conditionality and agency, what it is like sometimes to transcend that dualism, and what it can feel like to be more determined at times, and, at times, more profoundly empowered and responsible. Ortner's use of emotion helps affirm the qualitative existence of an *inner life*, or, at the very least, that a phrase such as "*inner life*" captures aspects of experience that evade other categorical descriptions, which emphasize behavior and bodily reactions.

This approach to subjectivity helps emphasize the irreducibly emotional character of experience and the inner life. Emotion is complex, however, and we should not understand it simply as sensation opposed to cognition, of feeling versus thought. It is visceral, in that emotion is possibly the quintessential experience of being a body, but this does not mean it is somehow separate from reason or cognition. As Martha

---

[8] As Luhrmann writes, "We know that our feelings are private, but we read them off each other's faces. We know that we know what we feel better than any observer, and yet we know that observers can see on our faces emotions we did not feel. We go to therapists to learn about the unacknowledged feelings that trip us up, and we believe in an honest, authentic emotional life" (Luhrmann, "Subjectivity," 349).

[9] Luhrmann, "Subjectivity," 345.

[10] Ortner, "Subjectivity and Cultural Critique," 31, 34; Luhrmann "Subjectivity," 345, 347.

Nussbaum has argued, emotions need to be understood as forms of knowledge that tell us about ourselves, our world, and situations in which we are involved.[11] Referring to compassion, for example, Nussbaum argues that this emotion is itself a "certain sort of reasoning." It is more powerful because it is not only evaluative but also has the force of emotional urgency.[12] Emotions help us in appraisal and always come with content that tells us something about our engagement with the world. Emotions such as guilt and regret, frustration and despair, as well as hope, gratitude, relief, and elation have the potential to be self-revealing and tell us something about our world and our engagement with it. Just as pain as such can tell us there is something wrong with the functioning of the organism, more complex emotions can signal changes in relationships, lived patterns, or one's identity.

We can think of emotions, then, as knowledge or sensational responses with epistemic import. They are felt, are bodily, and bring a more holistic vision of knowledge located not just in the mind but also throughout one's physical organism. We can even go further and, with Arthur Kleinman, think of the body, and so also emotions, as the interface between what we experience as interior and what we experience as outer, creating greater emphasis on the role of the emotions, affect, and embodiment in moral transformation.[13] Kleinman's work, indeed, further emphasizes emotion as signifying a dimension of inner life, as emotion reflects the confluence of various phenomena that we often categorize as inner or outer. This makes the emotional life central to our understanding of subjectivity, affirming the subject as a complex nexus of agency and conditionality, inner and outer, and providing a rich take on cognition, knowledge, and relationality.

This understanding of emotion, desire, and other facets of the inner life is something that we will take up again in the following chapters when discussing Iris Murdoch's metaphysics and psychology. For now, such an approach can help us sketch a definition of subjectivity that is complex and rich, with a depth that extends far into the world and deep into the body.

---

[11] Nussbaum, *The Fragility of Goodness*, 45. See, for example, Nussbaum's *Therapy of Desire* for more extended discussions.

[12] Nussbaum, "Compassion," 28.

[13] Kleinman, "How Bodies Remember." This understanding of the body is the key social interface is anticipated by Merleau-Ponty's understanding of the body as a subject through which the world is made meaningful (Merleau-Ponty, *The Phenomenology of Perception*, 75–99).

It is a conception that Ortner sums up in this way: "By subjectivity I will mean the ensemble of modes of perception, affect, thought, desire, fear, and so forth that animate acting subjects. But I always mean as well the cultural and social formations that shape, organize, and provoke those modes of affect, thought, and so on."[14] This account represents, then, the human person as a lived, felt phenomenon whose agency is real yet shot through with social, cultural, and political influences.

Such an understanding of subjectivity makes it highly relevant for a project like this one that hopes to account for the way violent political upheavals transform the individual. The emotional aspect that Ortner emphasizes, for example, as further developed by Luhrmann, allows us to view the emotions that can arise from moral harm—such as anger, fear, guilt, shame, lack of empathy, and despair—as communicative. When we apply such notions of emotion and subjectivity to situations of political violence, we can understand such emotions to have epistemic content about violence that, upon reflection and with the proper resources, tell us and survivors of political violence something about the experience of violence and moral development.

We can see this if we look back to the Dizdarević quotation that opened this work. We would be hard-pressed to understand his words if we viewed the self as an essentialized or atomized individual. Even if we thought of the self as malleable to an extent, we cannot do justice to Dizdarević's insights if we think of the self as a personality or character that life's trials do not affect to its core. Dizdarević is stating that "civilization" has been completely undone, and with surprising swiftness, by the violation not so much of bodily integrity, though that is represented in his words, but more specifically by the violation of human dignity, of neighbor betraying neighbor, betraying something we could call almost sacred. It is a violation of trust, of community, and of one's moral standing and self-regard through bodily injury. It is a violence that starts with, but ultimately targets something deeper than, the skin.

This is not, then, just the collapse of some abstract concept called "civilization." Dizdarević gestures toward a collapse of the individual when he speaks of it in terms of a "distress that cannot be forgotten." It is emotional, visceral, and permanent. It "removes all finer human sentiments," transforming how the individual feels and how she will react through the affections toward another human. And it "wipes out any sense of justice,

---

[14]Ortner, "Subjectivity and Cultural Critique," 31.

compassion, and forgiveness." In other words, it transforms the virtues, transforms character—or at least the experience of oneself we call character and virtue—so that the person can no longer respond to others in the same way as she did before the transgression.

In a few short lines, Dizdarević's quotation reveals the way something as abstract as regional political conflict, itself conditioned by geopolitics (for instance, the end of the Cold War, in the case of Yugoslavia), can so affect one's local moral world, transforming relationships between neighbors, whose actions transform the subject. We cannot hope to understand such a situation, and how gestures like being hit by a rifle butt can undo one's worldview, without having a sophisticated understanding of the self that can account for activities on these different levels, as well as how the interaction between such levels conditions the self from moment to moment, breath to breath, year to year. There is more to Dizdarević's quotation, after all, than someone being struck with a rifle. There is the context of a fraying society, of institutions and symbols and bodies fought over and thrown down, that give that violent gesture its potent meaning. To quote Nussbaum again, "to grasp either love or tragedy by intellect is not sufficient for having real human knowledge of it."[15] All the more so, for extreme violence that tears the fabric of one's soul as it tears the fabric of one's community.

THE LIMITATIONS OF SUBJECTIVITY

Although Ortner and Luhrmann help us understand the ways in which emotion is a critical aspect of subjectivity, their frameworks are still not sufficient to explore adequately the moral dimensions of violence. Luhrmann argues that "subjectivity implies emotional experience" and that psychological theories of emotion can add clarity to anthropological approaches. In the course of her argument, Luhrmann seems to affirm morality in her schema, yet in effect, she ends up reducing it. She references the moral category toward the end of her argument to make some strong claims on the nature of emotion and morals. For example, she writes, "And yet that very basic, physiological, gut-based automatic judgment means that an emotional experience is a moral judgment ... Emotions are our most basic moral reactions."[16] This categorization seems to make

---

[15] Nussbaum, *The Fragility of Goodness*, 45.
[16] Luhrmann, "Subjectivity," 355.

emotions a form of morality, or at least a type categorized under the broader term of "moral judgment." The seeming consequence of this move is to make morality the more significant concept, as emotion is defined in terms of moral processes. Indeed, reading this charitably, Luhrmann's argument can be seen as a helpful formulation in that it gestures toward the evaluative aspect of emotion and the emotional dimension of morality.

The focus on emotion in her work, however, endangers the potential of this nuanced dynamism. "Moral judgment" in the above quotation is associated with "moral reactions," making moral judgment seem solely reactive, a result that elides other qualities of moral judgment and morality more generally, such as reflection, contestation, and negotiation in the moral life. Indeed, Luhrmann does not provide a definition of the moral, leaving absent an understanding of exactly how we are to understand it as a "visceral act" and how this connects with any notions of moral reflection and theorizing, which, although not the whole of morality, are ubiquitously human. Whereas Luhrmann defines emotion as comprised of six factors, for example, the moral dimension is mainly confined to one paragraph and becomes something of an auxiliary characteristic to a subject largely characterized in political and emotional terms. "If subjectivity is the emotional experience of a political subject," she writes, "then to articulate the psychological structure of the emotion only gives us more evidence to argue that power is inscribed upon our bodies and that moral judgment is a visceral act."[17] Although morality is certainly visceral and often comes from the gut, this formulation does not give an adequately complex understanding of moral experience. Where the moral seems at first to be the broader category, it comes to be defined almost entirely by Luhrmann's understanding of emotion, reducing its complexity.

I do not mean to reduce the role of emotions in moral discernment. As Nussbaum has noted, emotions are critical to evaluation, and, indeed, critical to self-awareness and moral growth. What occurs in Luhrmann's work, however, is a conceptual confusion about morality created in the very process of trying to create clarity around the role of emotion in anthropological approaches to subjectivity. The result is that, as the moral is muddied up with the emotional, the moral dimension is made much less necessary and is in danger of becoming unnecessary. Why even have categories of morality if morality is fully grounded in, created through, and

[17] Luhrmann, "Subjectivity," 359.

defined by the emotions? In the end, Luhrmann subsumes moral experience and dynamics within the broader, explanatory category of emotion. And, in the process of trying to affirm Ortner's move to expand notions of subjectivity to include emotions, Luhrmann ends up narrowing her understanding of subjectivity by reducing the moral to the emotional.[18]

Luhrmann's project is not explicitly one concerning moral development or morality, however, and it might seem uncharitable to try and place the goals of one inquiry on that of another. This is fair, but my reasons for spending time on Luhrmann's use of Ortner are twofold. The first is to show that an emphasis on other categories, such as emotions and the political, even when these encompass the moral, can too easily reduce the moral as a central dimension of subjectivity. In so doing, the moral dimension can be elided even as it is referenced. This can happen with any category, but as we saw in the introductory chapter, this is a regular lacuna in social-scientific and even humanistic inquiries.

Second, issues of subjectivity, including the way Ortner and Luhrmann frame subjectivity, have important moral aspects that must be kept in sight. Luhrmann defines the subject as a *political* subject, one "caught up in a world of violence, state authority and pain, the subject's distress under the authority of another."[19] These are moral issues and experiences, however else we may describe them, as they deal with the self immersed in coercive forces affecting one's subjectivity, shaping value, and transforming one's worldview. And if we are to take the testimony of survivors of political violence such as Dizdarević's and Maček's informants, we miss something if we do not emphasize the moral stakes that such political subjectivity involves.

The point is not to argue against an understanding of subjectivity that forefronts the emotions. It is to demonstrate, however, that even such a corrective move—one that even references morality—does not ensure that the resulting understanding of subjectivity will deal with the moral dimension of being a subject in a robust, rich manner, one in which morality and moral experience—really, the moral life—will not be reduced to other categories. What we require is an understanding of *moral* subjectivity that will frame the experience of being a subject, as well as related complex interactions with the world, in a way that forefronts the moral aspects of such experience.

---

[18] In structure, this is a similar move to those who would define the human subject mostly in terms of the will or reason, at the expense of emotion, but in Luhrmann's iteration, it is the moral that is the foil used to affirm the emotional.

[19] Luhrmann, "Subjectivity," 346.

Subjectivity and Local Moral Worlds

In addition to Ortner's and Luhrmann's components of subjectivity, we need further components to enrich our understanding of subjectivity developed so far and move forward to a conception of moral subjectivity.[20] Following Ortner's own approach, combining theorists from various theoretic schools, I turn now to a number of anthropologists mentioned briefly in the introduction who have worked on issues of political and quotidian violence and its effects on subjectivity. To create a more robust understanding of subjectivity, its experience, and context, it will help to draw on their understanding of subjectivity and intersubjectivity. Their work is relevant not only because they have dealt most directly with the ways in which political violence transforms subjectivity, but because they have also made strides toward understanding the moral dimension of being a subject. What will be considered are important understandings of what it is to be a moral subject, focusing on the concept of the *everyday*, of experience being inherently *local, and* the *fragility* of moral subjectivity. These writers also emphasize the political and the emotional, but, even more, they emphasize how local worlds and experience are critical to understanding the dynamics of subjectivity and violence, as well as gesturing toward the moral dimension of such issues.

Two overlapping dimensions to consider are the *everyday* and the *local*. The *everyday* is a frame critical to many anthropologists looking at violence and subjectivity and is important to think of in terms of the Bosnian War, because, as Maček has argued, her informants were concerned with

---

[20] Laidlaw has argued that practice theory approaches have failed in their attempt to balance agency and structure in social analysis. Theories such as Pierre Bourdieu's habitus end up interpreting every choice and action of an individual to the larger field and institutions that they are a part of. This is a helpful, cautionary read of practice theory, but his discussion of practice theory is a bit oversimplistic in that it does not acknowledge the way in which such theories can still be used in ways that see agency in a more robust manner. Indeed, Ortner, whom he criticizes, makes use of many approaches, such as Clifford Geertz's hermeneutic anthropology, as well as practice theory approaches. She is one to recognize many of the shortcomings of practice theory (cf., Ortner's chapter, "Updating Practice Theory" in *Anthropology and Social Theory*). Laidlaw does not acknowledge this adaptability, nor does he acknowledge the ways in which thinkers, such as Bourdieu, were originally received in ways that attributed a too-deterministic read to Bourdieu's approach (cf., Gorski, *Bourdieu and Historical Analysis*).

normality and the norms of daily life.[21] And Dizdarević's quotation is so striking, because he does not describe exceptionally violent actions. He is describing the actions of neighbors in one's own neighborhood and gestures that, although violent, are not extraordinary. The *everyday* is a term often employed not simply to describe a certain level of action and interaction, as opposed to more extraordinary events and other social levels, but also to basic forms of violence that are woven into quotidian life.[22] In addition to the *everyday*, the anthropologists discussed here focus on political violence and how it transforms the self by highlighting the effects of violence as experienced on the *local* level, the privileged site where larger dynamics, structures, and events—regional, national, and even geopolitical in scope—are directly experienced and incorporated into society.[23] Although both terms recognize the structural aspects of violence, and even its macro-level manifestations, they emphasize the eye-level, personal, and interpersonal elements of violence as it is experienced.

At the same time, they have an inextricably moral character as sites of moral development and change. Arthur Kleinman, in particular, is concerned with the central *moral* dimension of the local, as well as the way macro-level dynamics arise from such local networks and yet also affect, in return, everyday life.[24] Kleinman argues that such violence is situated in *local moral worlds*—networks and communities in which value is created and contested and in which evaluation occurs. Although the local in Kleinman's understanding includes economic relationships and other modalities of human thought, action, and organization, the moral dimension of experience and sociality are held up as key, as informants have related that such dimensions are central to what is truly important to

---

[21] See, for example, Veena Das, *Life and Words*, Arthur Kleinman, "The Violences of Everyday Life," Nancy Scheper-Hughes, *Death without Weeping*, Philippe Bourgois, "The Power of Violence in War and Peace," as well as Scheper-Hughes and Bourgois's edited volume, *Violence in War and Peace*.

[22] Scheper-Hughes makes a distinction between the more overt state violence and the indirect, everyday violence that comes from inequality, corruption, and social and economic dynamics and relationships, what Johann Galtung first described as *structural violence*. The main difference is that while Galtung emphasizes the impersonal nature of such violence, where there is no clear actor perpetrating a violation, Scheper-Hughes focuses on the experience of such violence as part of daily life (Galtung, "Violence, Peace, and Peace Research," 170; Scheper-Hughes, *Death Without Weeping*, 216–267 and *passim*).

[23] Kleinman, "How Bodies Remember." Scheper-Hughes, *Death without Weeping*.

[24] Kleinman, *Writing at the Margin*, 122–125.

them.[25] This turn to the local as the generative context of value and meaning also means that Kleinman privileges issues of subjectivity, intersubjectivity, experience, and social relationship. Further, local moral worlds are the reservoirs from which more explicit forms of violence, such as war, come out of and into which they are then absorbed. In other words, Kleinman's approach looks to the local to understand violence, with a keen eye to the relationship between different levels of social experience and interaction, whose dialectics help generate everyday violence and the subjectivities formed through such violence.

What Kleinman means by *local moral world* is one's immediate community or network. Kleinman writes that "'local worlds' is meant to emphasize the fact that ethnographic descriptions focus on micro-contexts of experience in villages, urban neighborhoods, work settings, households, and networks of bounded relationships in communities where everyday life is enacted and transacted."[26] It is moral, however, because such worlds and processes "consist of the contestations and compromises that actualize values both for collectives and for individuals."[27] One's local moral world is the very place of evaluation, care, commitment, and norms, as well as their creation, contestation, adoption, and elimination. Each individual, however, is complex, and has many identities and loyalties that are not reducible to just one identity. This also necessitates that we acknowledge the multiple moral worlds in which one lives. This makes *the local* implicitly plural, as well as moral, an umbrella term for the many local worlds one lives in, whether these are the home, school, work, a religious house, or other places that are the context of moral development and action.

Further, Kleinman's assumption is that it is on the local level where the individual and more macro social events and dynamics interact, are mediated, and influence each other. Macro-level structures and forces, such as economics and politics, are influential at the level of the local, that is, one's network of concern and care. This is the level where "what is most at stake for persons and families" is actually created, given form, relation, and meaning, and where the "sociosomatic linkage between symbol systems and the body, between ethos and the person" are created from contesta-

---

[25] Kleinman, "Pain and Resistance;" Maček, *Sarajevo Under Siege.*
[26] Kleinman, "Experience and Its Moral Modes," 358fn2; see also p. 359.
[27] Kleinman, "Moral Experience and Ethical Reflection," 71.

tion and negotiation.[28] It is also the place where the larger events of the world come into contact with value creation to transform what matters to people. In other words, a local moral world is one's immediately lived network where meaning is not only created but imbibed and embodied. This emphasis makes such networks inherently moral—indeed, the place of morality—in a double sense, as the level where one's values, concerns, goals, and visions of the good are formed, and also where they are affected by broader events and trends.

It is interesting to note that Kleinman developed his understanding of local moral worlds in the course of theorizing about chronic pain. The idea that the experience of pain is inescapably local is central to the development of this concept for medical anthropology.[29] Kleinman writes,

> And here, where persons encounter pain, is where we need to center the study of its sources and consequences. Thus, studying chronic pain patients means that each must be situated in a world. That world must be described, and the description must include an account of the experience of pain in the wider context of experience in family, workplace, and community. To understand what chronic pain signifies, what its experience is like, ethnographers must work out a background understanding of local knowledge and daily practices concerning the body and the self, and misfortune, suffering, and aspiration generally. And they must relate this background understanding to episodes of pain, courses of pain, and other aspects of the world of patients, families, and practitioners who are responding to the constraints of pain.[30]

The origin of Kleinman's understanding of local moral worlds in pain makes Kleinman's construction of *local moral worlds,* and of the *local,* particularly relevant to discussions of subjectivity and political violence, as demonstrated by his work in collaboration on such subjects.[31] It also shows that the use of similar concepts, such as *everyday violence,* are fundamental to all of these conceptualizations that aim to put experience and more intimate levels of interaction and life at the center of inquiry.

---

[28] Kleinman, *Writing from the Margin*, 123–124.

[29] Kleinman, "Pain and Resistance," 173. It should be noted that "Pain and Resistance" was included as a chapter in Kleinman's *Writing at the Margin*, from which I draw the above quotation. I draw from both sources throughout.

[30] Kleinman, *Writing at the Margin*, 124–125.

[31] Das, et al., *Social Suffering*; Das, et al., *Violence and Subjectivity*.

What is shared by each of these writers is an understanding of the fragility, the vulnerability, not only of the moral but of the social as well. These sites in which one becomes, continues, and changes as a moral being are susceptible to change and even harm, making one's moral being and felt sense of character vulnerable as well. Veena Das, for example, writes how one's "access to context," as she calls it, and which I understand as one's ability to inhabit the geographies in which life and relationships exist, may become hampered or lost.[32] Das argues that trust in the durability of one's own context and relationships may be lost, swept aside by violent political upheaval. Such betrayals can create a loss of the ability to inhabit one's world—including the linguistic resources of that world—lessening abilities and avenues for response, until it can seem that the only response possible, other than violence and blame, exile and escape, is a form of lamentation, where one learns to live through the aftermath of violence by mourning, and so acknowledging, the immensity of the loss.[33]

Specifically, Das emphasizes the important epistemological dimension of this experience of violence. She labels as "*poisonous knowledge*" the survivor's experience of betrayal by family members and how easily loyalties and obligations that she once took for granted succumbed to the pressures of violent crisis and failed her in her hour of need. This knowledge can lead to frustration, regret, and even a feeling that the givenness of one's worldview can be thrown into doubt. This poisonous knowledge can itself lead to further violence if not dealt with correctly. It is a powerful epistemological transformation that one must either lament, and so begin moving through to a new worldview, or respond to with violence, in an attempt to refute such knowledge and blame others for the collapse of one's world.[34] Das builds on Kleinman's notion of local moral worlds and the relationship between different social levels with an account, both

---

[32] Das's understanding of context shifts in her work and, like other important terms such as *ordinary*, is not explicitly defined. Context does seem to imply one's social context of meaning and orientation, which connects to references of context as *lifeworld*. Das also refers to a "linguistic context," which is important to her work, as she is looking not only at language philosophy but also at how articulation is wrapped up in the experience of violence (Das, *Life and Words*, 9, 65).

[33] Das, *Life and Words*, 6–9.

[34] Das, *Life and Words*, 54–55, 77, 240fn29.

epistemological and psychological, that further defines the contours of how violence and social change can transform subjectivity.[35]

Experience, then, is privileged, although it is not the final authority, as reflection and analysis are needed to bring together the differing levels and dynamics at play. These emphases—the everyday, social vulnerability, the location of the moral in the local, and experience—are all methodological assumptions that I argue are critical if we are to understand the moral experience that individuals such as Maček's informants claim, and if we are to understand the moral stakes their experience bears witness to. Together, they begin to create an understanding, really an approach, to political violence that is better able to take into account individual moral experience, and so better hear, if not entirely articulate and account for, the claims that arise from such experience.

In the following chapters in which I engage with Iris Murdoch's thought, I will return to some of these concepts and show how Murdoch can help deepen their meaning, as well as show how they can help frame more fully some of Murdoch's concepts about the moral life.

Concepts such as the everyday and local moral worlds can also finally help round out the conception of experience I sketched earlier. The everyday and local moral worlds are the locations where experience occurs and through which actions and events on more micro levels, whether regional,

---

[35] It is important to note that although Das uses the term "*local*," she uses "*the ordinary*" much more often, and in a more technical sense, in *Life and Words*. Her question is how one is able to return to ordinary life, now devastated by life, and to move forward after the devastation (see also Das, "Ordinary Ethics"). This is very similar to the *everyday* of Scheper-Hughes, although Scheper-Hughes and even Kleinman, in his discussion of the local, seem to emphasize more the structural violence of the *everyday* in a way that Das does not. This does not mean that Das ignores such a dimension of routine daily life. Instead, the difference is one of emphasis. It also has to do with the different locations of their ethnographic fieldwork and how the specific cultural textures of each location affect the terminology and the meaning of such terminology in the anthropologist's explication of events. Das's *ordinary* holds the cultural resources for dealing with violent events, and she emphasizes how women in particular live into certain cultural roles as a way of acknowledging their suffering while moving forward in life. *The ordinary* and *local moral worlds*, then, do not exactly overlap, as they have a difference in scale. *The ordinary* seems to be expansive of life, except for violent events, while *local moral worlds* could be said to be the contexts and matrixes of relationships and norms that make up what we experience as *the ordinary*. In effect, however, Das and Kleinman seem to be after the same thing, that is, to account for the way one's life is localized and how what we consider to be violent events or periods interact with our everyday lives, as well as what about such interactions and our local worlds enables continued social life after explicit violence.

national, or international, are mediated in relation to the individual and more intimate, interpersonal levels of common life. They also bring out the moral dimension to experience, as these are also the locations in which moral development and change occur, and where evaluation and learning to evaluate also occur. In defining experience in this way, I intentionally design the frame I am constructing to privilege the individual eye-view, while trying to include other, more macro aspects of the moral life. This attends to but refocuses attention from the economic, political, and social to the moral, defined intimately, bodily, and interpersonally. In doing so, it creates an interpretive frame that can address the effects of violence on the individual in the context of a wider world.

Subjectivity to Moral Subjectivity

Even with a focus on subjectivity in studies of political violence, however, the moral can still be elided. Unlike Ortner and Luhrmann, where it can become subsumed under emotion or obfuscated through a lack of clear definition of what comprises *the moral*, anthropologists such as Kleinman focus their work particularly on moral issues, going so far as to create frameworks, such as *local moral worlds*, meant to reframe social-scientific research to engage first and foremost the local, subjective experience of moral loss. One can get a sense from this description how central Kleinman has been in the effort to reorient ethnographies to attend to the inherent moral dimension of experience. Yet, even with Kleinman, there is a limit to how well his framework can actually give voice to those experiencing violence and help us understand exactly what is occurring when a survivor of violence claims to have lost the ability to be "good." Even though he appeals to the term "*moral*" and focuses on experience as inherently moral, he lacks the vocabulary—we might specifically say the second-order vocabulary—to build on this and actually flesh out what such experience is like.

These limits become apparent when Kleinman discusses specific cases. For example, in laying out his idea of experience in the article "Everything that Really Matters," he uses research conducted with his wife and collaborator Joan Kleinman on the psychosomatic illness of Chinese individuals who had survived the upheaval of the Cultural Revolution. The Kleinmans draw on their informants' specific bodily complaints to better illustrate to the reader how symptoms and emotions that may seem straightforwardly medical in nature, such as dizziness and exhaustion, can actually be evidence of intersubjective dynamics and be a response to political realities and pressures. His informants, traumatized by the Cultural Revolution, were not permitted by the Communist Party or the state to criticize that period of history, and so they could not express the profound trauma of those times on their lives and the lives of their loved ones. The Kleinmans describe times when some of these individuals came together in small groups, and, instead of politically critiquing the state from the basis of their experience, their individual traumas would manifest themselves as a shared medical trauma. On a broader scale, the Kleinmans argue that suppression of expression turned to physical ailments, which in their aggregate affected society at large, showing how unjust silence could lead to social malaise.

For example, one such informant might be severely depressed from the fact that she could not express what she had experienced and felt, a suppressed suffering that would manifest as *dizziness* and *exhaustion*, perhaps a kind of hermeneutical injustice.[36] As the Cultural Revolution affected the entirety of the country, such feelings had a collective effect and manifested in the body politic, resulting in the "devitalization of social institutions and networks." The Kleinmans argue that such symptoms may reflect medical needs, but, more importantly, they also reflect political upheavals, the experience of which such individuals could not easily talk about in the open due to the authoritarian nature of their government. This is a fascinating read of symptomatology and its connection to political life, yet in a work looking at the importance of the local and the moral—and of the frame, *local moral worlds*—Kleinman does not directly discuss such symptoms as possible indicators of changed moral subjectivity or ability. His theoretical vocabulary, so good at illustrating how different areas of human life and experience are interconnected, as well as the somatic effects of political upheaval, lacks terms that could bring out the moral contours of that experience.

Kleinman has at his disposal terms such as "*moral*" and "*concern*" but he lacks the more elaborate and elaborated-upon vocabulary found in ethics or philosophy referring to types of morality, to moral development, ends, justice, teleology, character, habit, disposition, and goods. Indeed, we have a description of one's experience as dizzying, exhausting, devitalizing, but how did this affect one morally? When the individuals come together within a local moral world, how was the sense of their own morality changed? How did the Cultural Revolution change their sense of

---

[36] Fricker, *Epistemic Injustice*.

goodness and their relationship to it? And if they changed, what in them changed that resulted in a transformed moral self?

I am tempted to say that these questions are simply not Kleinman's and that putting them critically to his article is not only less than charitable but a misreading of a researcher's intentions. I resist that temptation, however, because of the stress that Kleinman himself has placed on the "inherently moral" nature of political experience. Indeed, in this article, Kleinman describes his informants' experiences as relational in nature, occurring in "a family, a network, a neighborhood, a community." Such relationships are at the heart of what makes one's morality and are the elements that constitute one's *local moral worlds*. Participation in the life of family, of community, is to participate in and eventually internalize the creation, negotiation, and contestation of values and what matters in life. Yet, the moral dimension of this experience is left unexplored and unarticulated.

What this should make clear is that an interpretive frame that can adequately give voice to, and provide an account of, the form of moral loss experienced by those involved in war needs a robust vocabulary as well as a rich conception of the human person. Such a conception needs an understanding of social and cultural conditioning, as well as how radically dependent individuals are on context and the broader events in the world. It also needs a way to describe one's inner life and the subjective experience of having one's moral architecture changed or, at least, experiencing it as changing. What we need, in order to better ensure that we include the moral in our understanding of the self, is a *moral* subjectivity, a way of understanding human experience that at once forefronts the social nature of being a subject, the political nature, yet also captures the interiority of such experience. What I mean by interiority is that we do not lose sight of the individual's personal world of feeling, thought, and value even as we stress how one's "inner world" is part of a larger fabric running through one's relationships and one's community. This is no easy balance, and both the social sciences and humanities have struggled to represent this complexity, which is at the heart of shared human life.

An understanding of subjectivity will be truly *moral*—a *moral* subjectivity—if it succeeds in expressing a vision of the individual as one for whom life is saturated with moral meaning and can be best described as an ongoing process of moral or virtue development among and in relationship to others.[37] Prior to conceiving of the human as primarily a center of reason, of will, of autonomous choice, a moral subjectivity represents the human as one for whom, to paraphrase from one virtue tradition, the questions of who am I and whom should I be are existentially central.[38] "Moral" here means not just issues of right and wrong but, as Charles Taylor might articulate it, can be defined more robustly as an orientation to what makes life worth living.[39] I would expand on this somewhat and describe moral as an orientation toward goodness, however that is understood, and the development that enables one to live a life aimed at goodness, embodied in particular images of the human and other horizons, and directed away from the vicious and unvirtuous images of the human. To develop morally, then, is to grow in one's ability to live into certain images, to orient oneself toward the good, as well as to develop one's capacity to refine and even change such images to reflect one's changing experience.

Moral subjectivity is meant to reflect, as Bernard Williams has said about Socrates, the primary concern of individuals over who they are, who they want to be, and how to be in the world and toward others.[40] It is an understanding of the moral dimension of experience, what is often called the *"moral life,"* that is not reduced to a code or rules but rather focuses on orientation, images and symbols, narratives, horizons, gestures, meaning, and embodied practice that create a life and one's understanding of life and its meaning. Identity, then, is central to this formulation of the self, uniting moral aspects of the self with issues of identity, as the question

[37] I define *moral subjectivity* in a particular way that is different from some others' uses. It is a term that is used yet not often defined. In philosophy, for instance, moral subjectivity can refer to questions that arise around the moral subject. These include issues of moral relativism involving individual moral viewpoints, as well as universals in general, such as a Kantian understanding of impartial reasoning (Meyers, *Subjection and Subjectivity*, ch. 1). It can also be defined quite broadly as "a moral position or intellectual space shaped by, but also constituting or shaping, discourse and material practices," making it akin to *habitus*, but more explicitly moral and personal (Reitan, *Making a Moral Society*, xiv). Uses akin to the latter range from those that seem to reflect a sense of a fluid, temporary way of being conditioned by larger forces in society and the world to understandings that involve a program of moral development, such as monastic training (Zigon, "On Love;" see also Berkwitz, "History and Gratitude in Theravada Buddhism").

[38] Williams, *Ethics and the Limits of Philosophy*, 1.

[39] Taylor, *Sources of the Self*, 4.

[40] Williams, *Ethics and the Limits of Philosophy*, 1.

"Who am I?" implies, as is a metaphysics or worldview in which such an identity and the answers to these questions will be comprehensible.[41]

To systematize this, we can say that there are six key elements, already discussed previously, needed to make a subjectivity primarily moral. These are, first, the subjective experience of what matters to one (values). Such value will differ from person to person, from geography to geography, from culture to culture. Overlap is possible, and possibly common. As each self is a particular center of differing cultures and experiences, however, it will always be somewhat unique. Second are the institutions and structures and their internalization by the individual that condition the moral imagination and what is and is not possible, particularly concerning how one is to be in the world (ethos). Third is one's understanding of how the world and society are and how the world should be (worldview), making it an area of life that is both evaluative and prescriptive or aspirational. This includes, fourth, the conditions and practices that make up one's worldview (practices). Ethos and worldview are practiced and internalized through specific acts in differing institutional contexts (school, work, family, etc.).[42] Fifth is moral development, of aspiration and failure, hope and despair that together with the previous constituents form the moral subject. Sixth, in addition, is an understanding of moral subjectivity rooted in local moral worlds and the maintenance of such worlds. This is the location of experience—the local—and through which more macro events and trends are felt, observed, evaluated, and mediated.

All of these are experienced as deeply intertwined and are separated only analytically in order to highlight important influences and constituents of being a moral subject in the midst of other moral subjects and their shared life, institutions, and structures (moral intersubjectivity). They interact and help condition each other. For example, one's ethos and worldview come from practices and interactions and contestations in local moral worlds, and yet such worldviews also have their own power to affect

---

[41] Taylor, *Sources of the Self*, 27; Geertz, *The Interpretation of Cultures*, 126–7.

[42] My understanding of ethos and worldview I take from Clifford Geertz via Ortner (Geertz, *The Interpretation of Cultures*, 87–141). Geertz emphasizes that such practices and dispositions make worldview and ethos *seem* natural. This emphasis on seemingness or the appearance of naturalness is an opening for Ortner, I argue, to insist on the possibility that a Geertzian use of culture can also see culture as ideological and part of domination. Drawing on Raymond Williams, she is able to see culture as Geertzian, comprising "symbols and meanings, ethos and worldview," yet nevertheless also sees that culture is part of forces of domination and ideology (Ortner, "Subjectivity and Cultural Critique").

ongoing social life.[43] It is a complex web creating not only meaning but also subjectivity itself.[44]

Moral subjectivity as a concept, then, is important not because of an ontological or neurological statement it makes about the human individual. It is a term that highlights the moral dimension of experience and the complex ways in which one's feeling about their moral ability is effected significantly by various factors, many not under the individual's control. It is a term not meant to deny other important aspects of one's life, such as the emotional and political that Luhrmann and Ortner describe. Instead, using the frame of moral subjectivity privileges the moral dimension of one's life, as well as the moral effects that broader phenomenon, such as institutions, structures, and so on have on individual and collective life.

This sketch of moral subjectivity remains just that—a sketch. Or, perhaps better stated, it is the beginning of a frame that requires further refinement and explication. It is one thing to say that we need to look at a human life as first and foremost a moral one, but this remains a rather flat assertion without a vocabulary and images of the human to give an understanding of moral subjectivity in detail and to make it a frame that can be used to understand particular experiences. I have put forward certain elements of a moral subjectivity, but they remain general and without examples to bring to life what I mean by moral subjectivity. Just as Kleinman could use *moral* and still not give a robust account of that experience, a moral subjectivity requires a firmer grounding in a more specific conception of the self with a rich, descriptive vocabulary to make such a conception effective. In the following chapters, then, I engage a particular notion of virtue and metaphysics, that of Iris Murdoch, to deepen the general understanding of moral subjectivity for which I have argued.

[43] Weber, "The Social Psychology of the World Religions," 280.
[44] Geertz, *The Interpretation of Cultures*, 5. This is a spin on Geertz's discussion of culture, where he wrote, "man is an animal suspended in webs of significance he himself has spun ..." I take this metaphor only so far here. Geertz, I would argue, does not believe in moral facts and signification in terms of the *Good*, as we will see in Murdoch.

CHAPTER 3

Moral Subjectivity and the Language of Virtue

An important lesson we can draw from the testimony of survivors of political violence is that there are identifiable, even unique, forms of vulnerability and suffering made possible when a person is caught in extreme violence and political upheaval. The moral life—the experience of one's self as a growing and changing moral subject among others over time—is said to enable flourishing. But it also displays vulnerabilities that can frustrate flourishing, as well as frustrate our projects, hopes, and dreams. As Rosalind Hursthouse has written, "…in evil times, life for most people is, or threatens to be, nasty, brutish, and short and *eudaimonia* is something that will be impossible until better times."[1] Such challenges are not new to formulators of virtue and virtue epistemologies, then, yet political violence presents forms of moral harm that seem to undermine the very fabric of one's moral subjectivity and felt character, sometimes eliminating the hope that such "better times" are even possible. If such experiences are approached with an understanding of moral subjectivity that was just presented, approaches to violence can, methodologically, be framed not just to take into account the formation of subjects but also to emphasize such formation—and reformation, even felt malformation—in its moral dimensions.

Dizdarević speaks of this loss in several entries in his war journal. His witness addresses the way in which violence can seem to transform individuals morally, even inverting their most cherished values. For example, he writes of one man so marked by the war that he savored the possibility

---

[1] Hursthouse, *On Virtue Ethics*, 177.

© The Author(s) 2019
J. Wiinikka-Lydon, *Moral Injury and the Promise of Virtue*,
https://doi.org/10.1007/978-3-030-32934-1_3

of vengeance, an experience he did not want to share with anyone other than those of his own community, who had suffered injustice as he had. The man was quoted in a newspaper saying, "We must not leave it to a stranger to fire the bullets that kill *chetniks* [Serbian and Bosnian-Serbian military and paramilitary forces and politicians]. I do not want to share that pleasure with anyone." It was not enough that his enemies die. They must die at the right hands. And such death was not a necessity but a "pleasure," the satisfying fulfillment of a desire. This sentiment goes beyond the instrumental, then, and the need to inflict losses on an opponent in order to gain a strategic upper hand. It does not even seem to reflect a desire for justice or balance, an eye for an eye, or to make sure wrongdoers receive what they deserve, or, at least, such a description does not fully capture the man's meaning. Instead, the man speaks in the language of sensuality, in the key of pleasure, the sensual and physical satisfaction of vengeance. This sentiment was so strong that he was actually saddened when his enemy died at the wrong hands. He spoke of his sadness when considering how a United Nations peacekeeper might kill one of the leaders of the Bosnian-Serb breakaway region: "Imagine how sad it would be if a Belgian or a Kenyan killed Radovan Karadžić or [Vojislav] Šešelj …," political leaders of the Bosnian-Serb breakaway region.[2]

What is arresting about this quotation is that it comes from a Sarajevan who, as Dizdarević writes, "built his whole life on the idea of tolerance, forgiveness, and a remarkably strong sense of attachment to his neighbors."[3] The war's violence and the experience of a neighbor's betrayal undermined those commitments, creating in him a new person, it would seem, who takes pleasure in the thought of killing those neighbors, a greed even that he would share with no other. Dizdarević does not reference this example to judge the man. Instead, it is given as further evidence that the war was destroying not just bodies and infrastructure but also the possibility of being a certain person in a certain world, of political violence itself as the destroyer of the conditions that make one's world possible. Indeed, in speaking of the possibility of forgiveness, Dizdarević writes, "But what we'll neither forgive nor forget is that they have broken what was best in us; they have taught us to hate. They made us become what we never were—and that is why, though they will be forgiven, we'll find it difficult to do so."[4]

---

[2] Dizdarević, *Sarajevo*, 33.
[3] Dizdarević, *Sarajevo*, 33.
[4] Dizdarević, *Sarajevo*, 34.

There is hope in such statements, even as there is an insistence on recognizing the transformative power of such violence. How, then, can we articulate and even account for such seeming change in character under the pressures of violence? I have already argued that we can see this violence as transforming one's moral subjectivity, the complex way in which one's moral life is constructed. Dizdarević is clearly speaking in a moral register, referring to forgiveness, betrayal, social coexistence, and moral horizons. Yet, to provide an articulation rich enough to do justice to the witness Dizdarević brings and to help articulate the experience of such transformation, what is required is a vocabulary and images of the human that can reflect the moral stakes involved. In other words, the concept of moral subjectivity I have sketched in a general and thin way that itself needs to be interpreted in a way that is expansive of such experience.

This is why I refer to the project undertaken in these pages as a *virtue hermeneutic*, as it is interpretive in a double sense. This virtue hermeneutic interprets a more general understanding of moral subjectivity to create a framework with the specificity necessary to analyze, articulate, and provide an account for certain experiences. To take a metaphor used in philosophy, anthropology, and religious studies, in a way it is to make such *thin* descriptions *thick*. This can create, then, an interpretive framework that can critically articulate and account for the ways violent devastation can result in moral loss. The double, although interrelated, goals, then, are to create an interpretation of moral subjectivity with the further, practical goal of being able to interpret real life experience. Such an interpretation, however, must be recognizable to those whose experience it. And it should also provide them with new insights into such experience, as well as a vocabulary to articulate the changes to their personality, identity, and moral ability that they have felt as a result of such violence. The application of such a framework should help provide the concepts and vocabulary necessary to clearly articulate that experience and also give an account of why such an experience seems so important and profound to individuals.

The *thickness*, to extend the metaphor, comes from particular notions of moral development, ability, and identity articulated through moral languages. Iris Murdoch's thought, which draws on virtue language to create an understanding of how humans grow and change morally, as well as the challenges to such development, is a rich basis for such a hermeneutic of moral subjectivity. The following pages provide an overview of Murdoch's thought that is particularly relevant to issues of violence and subjectivity, as there are other studies that speak to Murdoch's thought as a whole.

I pay particular attention to those aspects of Murdoch's thought that bear on moral development, as the harms that Dizdarević's and Maček's informants speak of touch on their ongoing ability to strive toward certain notions and images of the good. It is, in other words, a harm done to one's ongoing moral development. Building this understanding of moral development provides an account of how one's sense of their own morality is formed, sustained, and even weakened. This means that such an account begins to enrich our notion of moral subjectivity as something that changes and is vulnerable to various conditions, including extreme violence. Such an account, then, provides another step in arguing that moral injury is best seen as an issue of moral subjectivity, affecting the various conditions in which individuals negotiate their lives, their identity, and their everyday attempts at goodness.

Subjectivity and even *moral subjectivity*, however, are not terms that Murdoch uses; instead, she uses "*being.*" Beyond its derivation from Plato, however, who looms large in her thought, Murdoch never defines her usage of *being* explicitly. Indeed, even as she uses *being*, particularly *ethical being*, she emphasizes the processual character of the self, one that is conditioned by competing modalities of desire, political axiom, and duty, as well as various goods and loyalties.[5] Her use reflects an understanding of human existence as one of change, suffering, and love.[6] This is not the picture of the self as essential or resistant to an outside world as opposed to one of the interior, however. Indeed, even though Murdoch has been a champion for including notions of interiority as central to moral philosophy, her understanding of one's interior could only be understood as formed, and constantly in formation, by forces outside the self, as well as universal inclinations.

There are ways to read Murdoch, then, that make subjectivity an important, framing concept to bring out aspects, or at least potentials, in Murdoch's thought. Maria Antonaccio, a religious ethicist whose writings on Murdoch's philosophy have been seminal, has argued that Murdoch is best understood when placed within modern debates over moral subjectivity. Murdoch, Antonaccio writes, analyzed a tendency in philosophy to see the self as either too autonomous and solipsistic, which Murdoch called Kantian, or subsumed into a totalizing frame, which she called

---

[5] Murdoch, *Metaphysics as a Guide to Morals*, 492.
[6] Murdoch, *The Sovereignty of Good*, 50–52.

Hegelian. According to Antonaccio, Murdoch argues for an understanding of the self that mediates this position. The self is an integral being whose subjectivity is nonetheless tied to experiences and conditioning from the wider world.[7] This suggests that subjectivity is not only a helpful concept to understand Murdoch, but also, according to Antonaccio's reading, a critical piece in understanding the moves Murdoch makes. Murdoch sees the individual as more than a will but as a subject who embodies the tensions of being a location of competing influences and loyalties.[8] Using *subjectivity* to refer to Murdoch's concept of moral experience can help bring out the rich texture of her understanding of what it is to be a moral subject, as well as underscore the dynamic and oftentimes challenging fact of being a subject constituted by and in the world.

MORAL DEVELOPMENT IN MURDOCH'S THOUGHT

Philosophers who make extensive use of virtue language, particularly those labeled "virtue ethicists," have, despite this family resemblance, a range of views concerning human nature, the purpose of a generic human life, how easy or difficult it is to realize such a purpose, and the effect of adversity on character and moral development. Indeed, one of the classic (and classical) arguments many such philosophers have inherited from Attic-derived virtue ethics concerns how durable these traits called "virtues" are in the face of certain adverse events or conditions. Roughly speaking, there are two classical camps that engage this issue.[9] There is the Socratic-Platonic, where the virtuous man (and it is, sadly, always a man) can never be knocked from their virtue by chance or adversity, and the Aristotelian and neo-Aristotelian, which takes into account what has been called *moral luck*, where events can overtake even the most ethically realized individual and undo their happiness and flourishing.[10] A classic example is Priam

---

[7] Antonaccio, *Picturing the Human*, 6–12.

[8] "Freedom is not strictly the exercise of the will, but rather the experience of accurate vision, which, when this becomes appropriate, occasions action" (Murdoch, *The Sovereignty of Good*, 65).

[9] There are certainly others, including Stoic ethics, which, like Aristotelian and Platonic approaches, has a long shelf life in such discussions, although the Stoics come later. At least for some Stoics, like Epictetus or Seneca, the Socratic idea that the virtuous do not really suffer from misfortune holds.

[10] Nussbaum, *The Fragility of Goodness*, xiii–xiv.

from the *Iliad* who by all accounts lived a life anyone could envy. Priam was a good example of Aristotle's magnanimous man, the embodiment of the Aristotelian moral vision, until, that is, fate struck at the last hour of his life, undoing all of his happiness, reinterpreting his life from one of fortune to a long set-up for an unenviable fall.[11]

Murdoch is not easily assimilated to this binary, however. Although classical Attic philosophy certainly influences Murdoch, in particular her critical engagement with Plato, she does not share the Socratic-Platonic confidence in virtue, nor does she fully adopt Aristotelian and neo-Aristotelian teleology. This is to do largely with the influence not only of modern psychology, or at least psychoanalysis, but also of Christian theology's long and complicated reception and transformation of Attic thought. There is something like original sin to Murdoch's understanding of human nature, for example, an indelible pull against a life oriented toward goodness. As we will see, this is not framed in Augustine's idea of human will weakened from the Fall, but instead is framed in terms of vision and imagination. And even more than Aristotle's "moral luck," which sees in one's moral life a vulnerability to necessity and the material basis of human existence, Murdoch adds a more soteriological notion. Human beings are, perhaps tragically, forever trapped in their own egoism, and even if some of us can mitigate this to a significant degree, none can fully ameliorate this psychological, perceptual structure to human beings.

This has led Maria Antonaccio to call Murdoch's understanding of human life a pilgrimage.[12] Pilgrimage is a helpful metaphor, as it brings out the striving, effort, and the "pull," as she calls it, that accompany certain Christian notions of moral advancement. At the same time, as we will see, Murdoch is no theist, and although her thought is much more open to notions and metaphors of transcendence than most modern moral philosophies, there is no ultimate divine vision, no union, no taste of the divine on offer at the end of such a pilgrimage. This, despite

---

[11] There are, of course, other moral philosophers who use virtue language and do so without appeal to either of these traditions. For examples, see Hursthouse, "After Hume's Justice;" Driver, *Uneasy Virtue*; Slote, *Morals from Motives*; Swanton, "Outline of a Nietzschean Virtue Ethics." Moral philosophers grounded in ancient Attic philosophical systems have also begun turning to Confucianism to develop virtue ethics. See Slingerland, "The Situationist Critique and Early Confucian Virtue Ethics;" Yu, *The Ethics of Confucius and Aristotle*.

[12] Antonaccio, *A Philosophy to Live By*, 70.

the Platonic and neo-Platonic flavor in Christian theologies of spiritual ascent and sainthood.[13]

Murdoch's thought is also distinct from important contemporary neo-Aristotelian thought, such as Alasdair MacIntyre's idea of the moral life as a "quest." MacIntyre sees our proneness to disability and the vicissitudes of animal life on this planet as a limiting factor in achieving the end of a human life. For MacIntyre, we are all meant to become an "independent practical reasoner," his update on Aristotle's "magnanimous man."[14] Despite vulnerability and our "animality," as MacIntyre terms it, the goal remains and remains possible to achieve. For MacIntyre, who has also been influenced by the Christian reworking of Attic thought, and by Aquinas in particular, there are certain ends, certain potentials, that we as humans are meant to fulfill and realize. Life is a quest in that we need to see our life through cultural and religious narratives that reinterpret our day to day as a struggle for virtue and vice, with each individual the specific potential hero of such generic moral tales.[15] There is, of course, something dramatic in Murdoch's own work, of individuals working toward a more moral life against those inclinations to egoism structured into our very psyches. There is, however, no acknowledgment from Murdoch that there is an achievable *telos*, a final end to human life, that we can achieve. Indeed, to speak of our own heroism in moral achievement would be a way of reaffirming the kind of egoism Murdoch is so concerned about.

This is largely a result of Murdoch's view of human potential and psychology, which is less than sanguine when compared to some other virtue ethicists and moral philosophers, even MacIntyre. Murdoch was influenced by Christian notions of original sin, as well as by Freudian psychoanalysis, both of which claim there is a structural defect to human intention and motivation.[16] What this means for Murdoch is that there is no moral achievement or development that will ever deliver us to the far side of perfection. At the same time, this does not mean that images of perfection

[13] This is seen most famously in Augustine of Hippo's Neoplatonism but also in Pseudo-Dionysius. Good medieval examples include Dante's *Commedia,* as well as Bonaventure's influential *Itinerarium Mentis in Deum,* which he meant as a sketch of the stages of spiritual assent to divine union using Francis of Assisi as his model.

[14] MacIntyre, *Dependent Rational Animals, passim.*

[15] Interestingly, MacIntyre drops this notion of quest after his seminal *After Virtue.*

[16] It is interesting to note that original sin, as well as the psychoanalytic notions of egoism, were developed by authors (Augustine and Freud, respectively) who were both influenced by Platonism and Neoplatonism.

are not irrelevant. On the contrary, they are central to Murdoch's notion of moral development. Their centrality, however, does not lie in their achievability but rather in their instrumentality.

Murdoch's images of human *telos* are more in keeping with Reinhold Niebuhr's *impossible possibility*, an ideal serving as a standard that, although in reality unreachable, draws us toward it.[17] The vision of how one should be, although not projected as a potential to be realized, is still motivational in that it represents standards that spur us onward. Indeed, this is what Murdoch means when she speaks of the *Good*, which I will discuss later in the chapter, as "a distant transcendent perfection."[18] It reflects a view of human psychology and potential in which individuals are never beyond improvement. Each person strives to be good, but they are limited by a human psychology, a nature, that, to borrow from Christian soteriology, is fallen. In a pessimism about ultimate human potential that she adopts from Freud, yet also is saturated with theology, Murdoch argues that our motives are always mixed and that we are limited by an inherent selfishness that, although it can be diminished, can never be fully deracinated.[19] Although human beings are not Sisyphus, doomed to perpetual failure and disappointment, we will never be free from our efforts to push against a selfish nature toward a higher moral capacity. A moral life is possible, but moral perfection is not.

The danger for Murdoch, then, is our selfishness, or what she calls the "human propensity to *fantasy*." Fantasy is the innate disposition to view oneself as the center of all concerns. Human beings, for Murdoch, see themselves as the primary subjects in the world, marginalizing and objectifying (making into objects) the subjecthood, viewpoints, and even dignity of others. *Fantasy* is a good term as it affects our imagination and the way we see the world. Viewing all situations from the standpoint of a limited conception of subjectivity, where there is one main subject as opposed to many and multiple subjects, means that we do not see the world truly as it is. Our vision, our worldview, is a fantasy, at least to an

---

[17] Niebuhr, *An Interpretation of Christian Ethics*, 19. It should be noted that both Niebuhr and Murdoch share in the central theological concepts learned in youth, particularly elements of Reformed notions of the Fall, Niebuhr in his Unionist church and Murdoch in her Anglican, which includes Reformed influences.

[18] Murdoch, *The Sovereignty of Good*, 99.

[19] Murdoch, *The Sovereignty of Good*, 50.

extent, and through our selfish view of the world, we misperceive reality.[20] We see a world significantly different from the one we are in and significantly different from how others see it.[21] Humans embody, in other words, a Ptolemaic subjectivity with each individual at the center of her universe when in fact the world and subjectivity is more Copernican (at the very least). No individual has pride of place as does the sun. The rest of humanity and the world do not revolve around the individual, even though persons are so constituted that they have to work to make this perspectival shift from fantasy to a worldview more in line with our decentrality.[22] Another way to put this is that the self is the obstacle to seeing the relation between selves more clearly.

This leads us to Murdoch's understanding of moral development. By definition, such selfishness for Murdoch is the very opposite of goodness, of virtue, based as it is on a complete misunderstanding of the world and our place in it. Our needs, our desires, and our perspectives we see as normative. Since our orientation toward the world is quite literally a self-centered one, and since our problem is with how we perceive the world, ourselves, and others, the cure for such bewildered and bewildering souls lies in the virtuous cultivation of one's moral vision.[23] As Murdoch writes, "Serious reflection is ipso facto moral effort and involves a heightened sense of value and a vision of perfection."[24] Morality, or at least the moral life and moral development, is understood as the cultivation of the way we see ourselves and others, a cultivation that moves away from selfishness and toward a regard that Murdoch describes as loving and attentive.

Murdoch's thought is quite dynamic, then. There is an insistent pessimism of human nature that she does not want to undermine. She defines the baseline of human beings not so much by what one can achieve, although that remains ultimately important for Murdoch, but even more by the constant labor that must be undertaken for such achievement. Murdoch argues that the *Good* has a pull, just as innate egoism has a pull of its own, but this still requires effort, striving, to overcome some of

---

[20] This concept of fantasy brings together many of Murdoch's sources, including the fallen nature of Christian soteriology and the psychological viewpoint of Freud, as well as that of Plato, particularly the *allegory of the cave*, which Murdoch uses to illustrate that what we often see as real is little more than shadows (Murdoch, *The Sovereignty of Good*, 93).

[21] Murdoch, *Metaphysics as a Guide to Morals*, 342–348.

[22] Murdoch, *The Sovereignty of Good*, 50.

[23] Antonaccio, "Moral Change and the Magnetism of the Good," 144.

[24] Murdoch, *Metaphysics as a Guide to Morals*, 437.

fantasy's allure. This insistence is a constant remedy against reading Murdoch's talk of love and caring attention in too optimistic a light. As much as Murdoch focuses on her notion of the *Good*, much of her understanding of the self and the moral life is based on what must be corrected. Many of her central concepts such as *effort* and *attention* are defined as much as by what is wrong with human nature, a wrong that can be mitigated but not ameliorated, as by the human potential to resist and realize a degree of virtue.[25]

This should not cast Murdoch in too pessimistic a light, however. Though Murdoch's assessment of human psychology is critical, her reason for this criticality is to provide an account of moral development that shows how humans can become more virtuous and good. The cost for such hope, however, cannot be a moral philosophy ignorant of the brute facts of suffering, violence, and despair. Murdoch is no Pollyanna, and so does not pull punches when coming to her assessment of humanity. This does not mean, however, that Murdoch is therefore fatalistic. Despair must have its due for Murdoch, but this does not mean it must always have the final say. Although Murdoch does not think we can ever fully transcend our own psychology, there is always hope that we can transform the self, which is the seat of fantasy, into a self that is more caring, lovingly attentive, and that more fully embodies goodness in one's interactions with others and the world. Along with the obstacles, then, the human person also has potential to love and to see the world through a form of perception that is loving and attentive to the other.

Love, however, is not a Romantic term in Murdoch's thought but rather a technical, phenomenological one. The possibility of love in a selfish soul is the grounding of a Murdochian understanding of hope in human moral development. As Murdoch writes, "Love is the extremely difficult realization that something other than oneself is real."[26] It is a form of regard that allows one to decenter the ego and to value others—or at least to recognize that one *should* value others—as one more readily would value oneself. The key image of such hope and its potential is Plato's allegory of the cave. Murdoch argues that we need to develop in such a way that we orient ourselves away from the shadows, from fantasy, to the source of light, that sun shining provocatively at our backs. The shadows never disappear

---

[25] Murdoch, *Fire and the Sun*, 20. Murdoch also defines attention as "the effort to counteract such states of illusion" (Murdoch, *The Sovereignty of Good*, 36).

[26] Murdoch, "The Sublime and the Good," 215.

in such an image of the moral life, but one can learn in what direction one should cast one's sight and learn how to stare into that light for longer, sustained periods.

VISION AND ATTENTION

Vision is a key metaphor, because it is based on a number of factors important to Murdoch's philosophy, particularly the way that sight, as opposed to other traditional moral philosophies that, she argues, views morality as framed in terms of action and motion.[27] As a result, Murdoch argues that being good is grounded in the ability to have a more accurate moral vision. This is vision understood as the ability to see the world—really, to perceive and even imagine it insightfully—in such an attentive way that one's perception inspires a caring, loving response to others. Instead of using one's time looking at the small world of one's own concerns, which is quite an indulgence, one can practice turning, reorienting their senses to look beyond the small, cavelike world of their selfishness to the wide opening of the bright world that draws one out of their limited world and into the light. This is the ground of the classical Christian notion of becoming a convert, literally one who has turned around to see truth, one who has turned their back on falsehood.

Murdoch understands the content of such a moral conversion to be a new perception that sees the world and others with caring, loving attention. Indeed, attention is nearly synonymous with love for Murdoch. Caring attention is what eventually dethrones the ego by drawing one's consideration away from themselves. In its place is affection for others, drawing us to focus our attention on others and their needs, even if the ego will always tempt us, waiting in its cave.

How, though, does the right form of attention actually achieve this? What is the process? Although Murdoch begins her ethics and understanding of motivation with sight instead of action, her understanding of positive moral development constitutes a reorientation of perception that also creates a renewed ability to act. This process cultivates one's potential to be a better person who does not put their own needs and desires before others. We learn to do this through practices that hone our attention outward toward others and cultivating a perception outwardly oriented that is also charitable in character, we create a sympathetic, even empathetic,

---

[27] Murdoch, *The Sovereignty of Good*, 4–5.

view of subjects other than ourselves. Such practices of attention help develop an imaginative capacity to change one's worldview and ethos.

What is important to note is that, for Murdoch, loving attention is an orientation to the world that more accurately fits the facts of the world than does an orientation that is egocentric. Just as the world really does revolve around the sun, not vice versa, the self really is just one subject among a myriad, where each mote thinks erroneously that their concerns and viewpoint are central and of the highest significance compared to others.[28] One needs to develop a capacity to see others in a way that overcomes inappropriate self-love-as-egocentrism and directs that regard toward others. This is Murdoch's cure of the soul. This orientation empowers one to deal more accurately, and so more justly, with others and the world, as we are better able to see them not as less important than us but at the very least on par.

When it comes to Murdoch's understanding of *vision* and *attention*, however, there is some ambiguity. This is because there is a double sense in her understanding of vision.[29] This is important, as Murdoch, who is not a systematic writer, alternates her uses of key terms throughout her work. An example Murdoch uses—of someone suddenly seeing a kestrel—can help illuminate this. The kestrel's sudden appearance can grab one's attention away from selfish preoccupations and outward, away from the self, toward the beauty and truth of the world. She describes how such moments can change one's entire disposition. "I am looking out of my window in an anxious and resentful state of mind, oblivious of my surroundings, brooding perhaps on some damage done to my prestige. Then suddenly I observe a hovering kestrel. In a moment everything is altered. The brooding self with its hurt vanity had disappeared. There is nothing now but kestrel."[30] Seeing something so beautiful can snatch our attention and lead it outward to the world where we forget ourselves.

This example stresses the importance of the world as moral environment that demands our attention and asserts its presence upon us. Such

---

[28] Murdoch, *The Sovereignty of Good*, 65.

[29] Blum identifies five "visual metaphors" in Murdoch, which are "perceiving, looking, seeing, vision, and attention." He argues that they refer to different activities, yet Murdoch does not define their differences. Indeed, she can use them alternately, and their use is not consistent, as Blum acknowledges (Blum, "Visual Metaphors," 307, 309). This is another example of the inconsistencies found throughout Murdoch's works. They can be decried, but I prefer to enjoy the various creative interpretations that such inconsistency affords.

[30] Murdoch, *The Sovereignty of Good*, xii.

moments can call attention—really, grab our attention—and make us see how the reality of existence exceeds to an infinite degree the impoverished egocentric assumptions through which we see the world. The world has a moral pull on us, calling us to a *loving attention*, in Murdoch's words. And so we may think of attention as a type of moral perception, a mode or orientation that is itself moral, in that it is the capacity that allows one to pursue the good. Such loving vision draws us toward the *Good* because, at once, it pulls us away from egocentricity toward a focus on the subjectivity of others and even other things, lessening our tendency to fantasy.

The mechanics of loving vision, however, are also grounded more accurately in the moral reality of the world, as we all share these selfish dispositions and must struggle to be good, making compassion toward others not only more realistic but also more rational compared to an egocentric worldview. And here again we see that Murdoch's use of sight in her moral philosophy is quite literal, as she is arguing for the connection between the physicality of perception—actually sensing visually a small falcon—and the process of moral development. This magnetic characteristic of the world around us, and a characteristic of the human that allows us to be drawn out of ourselves and into the world, calls our attention to moral truths larger than our daily cares, concerns, and frustrations.

This is an understanding of vision understood as literal, but Murdoch also uses *sight* and *vision* metaphorically as well. She uses the metaphor of sight as a way to understand moral activity and development that pushes back against the metaphors of action and theories of ethics that stress discrete moments of judgment. To this extent, *sight* and *vision* and even *attention* are metaphors signifying not only physical processes but also the work of imagination in moral development. Moral vision, then, is also used at times to refer to a capacity of the imagination that can allow one to see another as a subject in their own right with their own dignity. Moral vision, in this understanding, describes a modality of imagination as a key moral faculty. Indeed, this may be *the* moral faculty for Murdoch. As already seen, the main culprit in immoral behavior is a tendency to fantasy, itself a work of the imagination. Reimagining ourselves as decentered, on the other hand, is virtuous. It is the imagination and the quality of our imaginations that are central to whether we are falling into our selfish predispositions or are working toward a potential that is loving and caring toward others and the world.

We can see what vision as a metaphor for the imagination might mean in practice by looking at a well-known example from Murdoch's *The*

Sovereignty of Good. It is the story of a mother-in-law (M) who did not think well of her daughter-in-law (D). In her account, D dies before M changes her opinion of her. As time passes and M grows, she thinks of D over the years and begins to think of D in a more loving light. She starts to wonder if she had been too quick to judge her daughter-in-law. Perhaps her judgments were based on her own snobbery and elitism and so were neither fair nor accurate.[31] The story shows the result of a change to M's moral imagination that allows her to reconsider her evaluations in the first place, yet it also shows how M continues to change morally beyond this in that she tries to think of D from a perspective other than that of her original more selfish attitude. What occurs is not a transformation of a relationship between two people. One member of the relationship is no longer present for that to happen. Instead, M herself transforms, seeing her former way of relating to her daughter-in-law in a more critical light. In the process, she becomes more caring toward D's memory and more attentive to the possibility that her perspective may have misled her in evaluating her son's wife.[32] She *sees* D, herself, and the world differently in her mind's eye. She does not do anything differently that anyone can observe, nor does her physical perception alter in any way. Her perception is a metaphor for her critical imaginative capacity that allows M not only to en*vision* D differently but also to reevaluate her own assumptions.

One can see, then, with no pun intended, the dynamic use Murdoch makes of vision and imagination in her philosophy. The way we see the world has much to do with the quality of our imagination. In the proceeding example, M could not look past her own worldview to consider that D may have had many attractive qualities. After all, her son, D's husband, was able to appreciate those qualities, presumably enough to marry her. M's limited worldview had a quite practical effect on how M lived in the world, how she related to this other person, and how she behaved. At the same time, after D passes, there is nothing left to see, yet M's imagination, her mind's eye if you will, does change, and she changes her valuation of the woman. This is an alteration not of D, who is no longer present, but rather of M.

This passage is usually understood as central to Murdoch's argument for interiority. The M and D example is largely meant to show that moral activity can occur within the individual without any outward indications,

---

[31] Murdoch, *The Sovereignty of Good*, 21–23.
[32] Murdoch, *The Sovereignty of Good*, 21–23.

and so combat overly simplistic theories of behaviorism in philosophy. As such, the example argues for interiority as an irreducible feature of human beings. Interiority for Murdoch is a form of subjectivity understood as having a "dimension of privacy, inwardness, and uniqueness that cannot be wholly reduced to its social, historical, and linguistic determinants."[33] It is a rich sense of the self that emphasizes one's actual experience of being an embodied subject, insisting that one's being is quite social, insisting on this without, however, reducing one's sense of interiority to other factors and influences.

The example of M and D, then, is quite rich and is reflective not only of Murdoch's battles against certain forms of behaviorism but of her larger sense of subjectivity as well. If we combine the story with the discussion of the kestrel, they point to a dynamic understanding of interiority as not separate from the world but already within it. The kestrel is *seen*, after all. It is through the visual sense that Murdoch's mental musings are confounded, changing her very pattern of thought. And, in reframing Murdoch's place in the world and what she values—moving from the minor slights she received to the beauty of the kestrel unencumbered by perceived humiliations—the moment with the kestrel helps Murdoch see herself and the world differently. The world acts directly through sight on Murdoch's imagination and worldview. Perhaps, too, M had a few "kestrels" in her life—occasions that altered her vision and subjectivity. Over time, this could have changed M enough so that the next time she considered D, she could do so differently and more charitably. Less centered on her own issues, she could possibly reimagine D in a way unclouded by past preoccupations and concerns that, with time, may no longer seem important. One can only imagine how this changes the way she would then both imagine and see the world and others around her.

This double movement in Murdoch—of emphasizing the moral nature of the senses while also using the senses as a metaphor for cognition—reflects the way that Murdoch understands the interior mechanics of moral development. There are some concepts—moral concepts—that have traditionally, Murdoch argues, been discussed metaphorically. This is not due to a poetic penchant on the part of philosophers of yore, but rather to the fact that such concepts are in their structure metaphorical and cannot be talked about in a way that assumes one can strip metaphor away to reveal the true nuts and bolts of an idea. In a way, this later way of thinking is

[33] Antonaccio, "Moral Change and the Magnetism of the Good," 146.

iconoclastic, itself an ideological position and no more "scientific" than that of an iconodule.[34] As children, and as adults, we learn ethics through images and narrative. Moral development is as much or more so imaginary as it is logical and the stuff of spare analysis. This undergirds Murdoch's understanding of morality and moral development as one mainly of sight and of imagination. The way we gaze is never neutral nor are the metaphors and images that make up our moral imaginaries.

These two examples are important, because one is more literal and the other metaphorical. As stated, Murdoch actually sees a kestrel. It is a sighting that seizes her attention and changes not only the moment but also Murdoch's mood at the time. Meanwhile, the M and D example is metaphorical. No one is actually seen, as D is dead. One can speak of the "mind's eye," but this is metaphorical, a way of articulating the work of the imagination as we ruminate over issues and images in our minds. This dual use of sight, as a metaphor but also description, which Murdoch never fully irons out, points to the very metaphorical nature of moral development for Murdoch. Sight is physical perception, but just like metaphors in ethics, one can't pull away the metaphorical layer to get at the more spare, mechanical truth.[35] We see the way we imagine and we imagine the way we see. They are part of the same, central process of moral development and how we engage with the world, one that reflects an understanding of mind that, although not reduced to a more simplistic behaviorism, is nevertheless a material understanding in which embodiment and morality necessitate one another.

Engagement here is key. Murdoch's thought prizes moral vision and imagination but as a part of a larger understanding of ethics as both being and acting in the world. Murdoch is known for her arguments for interiority, an example of which is the M and D thought experiment. This, however, can be emphasized too much. Murdoch is concerned with life and intersubjectivity, and imagination and sight are not just aspects of interiority but rather the direct engagement of a being in the world with the world. She pushes back against more action-oriented metaphors in ethics, but she does not do this to deny action and behavior in the moral life. Instead, she wants to create what she sees as a better grounding that connects a richer sense of the human person and moral development with how best we learn to be better moral actors and selves toward others. The

---

[34] Murdoch, *The Sovereignty of Good*, 75–6.
[35] Mulhall, "Constructing a Hall of Perfection," 221–22.

extent of our imaginative capacity to push back against our innate, dispositional selfishness allows us to envision the world (accurately, in Murdoch's view) as having countless subjects, and so to act not as the center but one being among a great many. Seeing this intersubjective world decenters us and so deflates our self-regard and how we value our own value positions. That practice creates a reformed imaginative capacity. Practice reveals the dialectic at the heart of Murdoch's moral development, one that is dynamic and takes account of perception, motivation, and action.

What does this say, however, about Murdochian virtue? For Murdoch, moral development consists of developing capacities and dispositions that can help convert one toward the *Good,* just as Plato argues we need to turn our backs to the shadows and turn toward the sun. The moral life is largely understood, in this way, as vision but also as orientation. The moral life, in other words, is moral development spread across one's life made up of a continuing pilgrimage to make more virtuous our imaginative and perceptive faculties so that we can constantly improve our discernment of the moral contours of the world and orient ourselves to live according to this new vision. This involves courage, temperance, and other classical and cardinal virtues, but also further virtues, such as creativity, openness, willing vulnerability, forms attention, and Murdoch's understanding love.[36]

What such virtues mean for Murdoch, however, she never explicitly defines, yet my sense of her understanding of virtue does not map on to a more Aristotelian understanding of *habit.* Certainly, Murdoch says in a very straightforward way that we are too selfish to have an "external point or *telos.*"[37] Murdoch does mention habit as being central to virtue, though she does this in passing and seems to assume that habit is so central to virtue that it does not need to be explicated.[38] This does not necessarily mean that habituation for Murdoch is connected to a *telos* in more traditional Aristotelian understandings of virtue. Although not without a sense of habituation as part of moral development, I take from the preceding discussion that Murdochian virtue is better understood as those dispositions and capacities, with an emphasis on the physical and metaphorical dynamics of vision and attention, that allow one to orient oneself toward the good. It is true that, as a part of this, Murdoch presupposes the many classical virtues such as temperance and courage, as well as more

---

[36] Murdoch, *Metaphysics as a Guide to Morals,* 86.
[37] Murdoch, *The Sovereignty of Good,* 76.
[38] Murdoch, *The Sovereignty of Good,* 89.

Christian-influenced concepts of love and care. She also sees virtue as accumulating, even building one's propensity to act against selfishness.

At the same time, Murdoch sees human behavior as constantly drawn down by *gravity*, a term she interprets from Simone Weil's writings. There is always a default disposition of fantasy, vice, selfishness, and egotism, even if one develops a counterbalance of care and decenteredness. Murdochian virtue, then, corrects and mitigates, but does not seem to ameliorate, what is a key tendency in human moral life. This differs from more traditional virtue accounts where one may be able to look back at the end of a life and call it excellent, or even Christian (particularly Thomist) theologies, where one, despite imperfections, may still gain the beatific vision. Practice brings virtue, but it is not guaranteed, nor guaranteed to remain.

MURDOCH'S METAPHYSICS

I have so far referred to the *Good* several times without explaining how it functions in Murdoch's thought. This needs to be remedied, as it is central to her metaphysics, which is in turn central to her moral theory and what Maria Antonaccio has called Murdoch's *realism*. Murdoch's first, best-known work in philosophy, *The Sovereignty of Good,* puts this concept literally at the front of the work, thus signaling its importance. For Murdoch, the world insists that there is something like *moral facts* that call for our acknowledgment.[39] This brings us back to the previous discussion of loving attention as being more accurate. In the act of attention, individuals are not just treating others charitably; they are acting out of a realization of the fact that we are less important than we think we are. This for Murdoch is a fact of human life and experience. Focusing on others instead of ourselves is an action that is much more in line with social reality than self-absorption.

These moral facts, however, are not discerned through a universal reason applicable to all, nor through a cost-benefit evaluation of outcomes. Instead, the moral life is the development of the self into a being whose dispositions, moral imagination, and moral perception more accurately reflect one's place in the world and the needs of others. This empowers one to be a better, more virtuous person and to respond justly to other

---

[39] Murdoch, "Vision and Choice in Morality," 95; Broackes, "Introduction," 1.

subjects.[40] Foregrounding any discreet moral action, then, is the ability to see the world, to see others and ourselves, and our relationships and situations as accurately as possible. One's quality of moral vision, not one's purity of will or reason, determines not only one's ability to understand the best courses of action, but also, much more broadly, the best way to be in the world. The struggle in moral development, then, is not so much a matter of will, simply understood, but in our propensity to see ourselves as the measure of all judgments and concerns, instead of seeing others as inhabiting a subjectivity equally deserving of care and dignity. The struggle is to cultivate vision that will empower us to see the world in a way that will counter that propensity so that one can orient oneself toward the *Good* so perceived.

The idea of something akin to *moral facts* assumes a metaphysics, one that points to a broader understanding of what constitutes the moral or ethical. This is where Murdoch's understanding of the *Good* comes in. There is a combined emphasis in Murdoch's thought on, first, being able to see and recognize what is good as opposed to what is generated by and through our selfishness, and, second, developing the ability to orient our life, thought, and actions to reflect that *Good*. Growing as a virtuous or good person requires the ability to see this *Good* more clearly and also to orient the structure of one's life in accordance with such perception. The *Good* as a metaphysic, then, represents the ground and justification of morality. What, however, is the *Good*?

Murdoch's understanding of the content of the *Good* in her writings remains curiously obscure.[41] It reflects a Platonic notion of an objective moral domain, and Murdoch's writing shows its multifaceted debt to Platonic thought. This suggests that one could turn to Plato to supply the meaning and content of this *Good*, yet Murdoch is not slavish in her use of Plato. She spent an entire, early work, *The Fire and the Sun*, wrestling with

---

[40] Murdoch, *The Sovereignty of Good*, 91–2.

[41] One cannot be expected to perceive the *Good* or the world as it is firsthand. What this means is that it is impossible to describe the *Good* in full. Some will find this formulation frustrating. Indeed, literary theorist and critic Terry Eagleton expressed such frustration when he wrote, "There is, however, a price to be paid for this breathtaking generosity of vision. *Metaphysics as a Guide to Morals* (the title surely parodies *Zen for Business Executives*) is a rambling, repetitive ragbag of a book, the philosophical equivalent of Murdoch's devotion to the loose baggy monster of a novel. It sacrifices rigor of thought to imaginative scope, and some of its more technical sections have a generalized, second-hand feel about them" (Eagleton, *Figures of Dissent*, 259–60).

Plato and arguing against some of his conclusions, particularly about the role of art.[42] Murdoch also draws on various sources, from Freud to Zen Buddhism, to develop her philosophy, making such direct correlations too simplistic for her thought.

We are left, then, with questions about this *Good*, yet this has to do not so much with an omission on Murdoch's part but with the nature of the *Good* itself. Each individual is constantly growing into their ability to discern the *Good*, and Murdoch's job as a philosopher is not to provide a definitive account of its contents, as if she were a prophet or sage revealing a hidden scripture. Understanding the *Good*—if "understanding" is even the correct term; perhaps "*inhabiting*" or even "*successfully seeking*" is better—comes not through ratiocination alone but rather through life-long moral development. If Murdoch placed the philosopher in the privileged position as having access to the *Good* as such and in total, it would necessarily result in a philosopher who was somehow inhuman. It would imply that the philosopher had fully grasped and even inhabited the *Good*, and, as a result, saw all things clearly and was in need of no further development. Murdoch's element of pessimism rules this out.

The *Good* is not, then, best understood as a set of prescriptions or a description that we read and enact. It is, instead, something we intuit, employing not only imagination and thought but also bodily organs, such as the eyes. This conception of orienting oneself toward the world, then, is much more embodied than it is intellectual or idealist, and so it would not be possible for someone to present the content of the *Good* in treatise form without missing a good deal of it. This would mislead the reader and miss the philosophical mark. Instead, the *Good* is a transcendent perfection that propels us forward, like an impossible possibility, as our appreciation for goodness, defined as loving and not egocentric (how the world really is), develops as we grow and learn, and, hopefully, as we realize how far we have come and how far we are yet to go. It is not comparative, in that what the *Good* is changes as we change. Instead, its perfection of perfectly being the *unself* will, if allowed, constantly drive us to compare how we are and how we imagine our world to that of the *Good*.[43]

The *Good* is also to be understood through one's interaction with the world. Again, Murdoch puts emphasis on the imagination, yet this might

---

[42] Published as a separate monograph, *The Fire and the Sun* is included in full in the collection of much of Murdoch's work, *Existentialists and Mystics*.

[43] Broackes, "Introduction," 73.

seem to be in tension with her realism that focuses on a more objective standard. This is not something Murdoch systematizes. It does not mean, however, that Murdoch understands the individual to be a self who freely choses the truth or perceives the *Good* free of history. The world will make claims on us, as philosopher Heather Widdows points out, so that we cannot say that we invent or imagine value. Imagination is not understood in this way as unbounded creativity. *Fantasy* as a concept is a bulwark against this interpretation. Murdoch seems to be insisting, on the contrary, that we use all of our faculties, and particularly our imagination, to discern value that is already there, or, at the very least, struggle toward this goal with the merited assumption that the world is not morally neutral. For Murdoch, evaluation is not just a matter of unencumbered choice. Instead, we need a metaphysics that will help stress the fact that we are already always moral creatures who find ourselves in a world already saturated with meaning and value.[44] We see, we interpret, but we do so as already conditioned, already moral beings, within a world already populated with other moral beings and stakes, making it a moral landscape. Indeed, sight and imagination imply that we are engaging with things already present, instead of willing them or acting upon them, although poor sight or imagination that wanders into daydreams can perceive the world unclearly. For this reason, Antonaccio calls Murdoch's philosophy a type of realism in that it stresses the world to be already full of value, one to be engaged and sensed, not one to be conjured.[45]

Murdoch's metaphysics is not without tensions, of course. This would be true, even if she were more systematic. The point of covering her metaphysics here is to provide enough of an understanding of Murdoch's main concepts to then create a practical framework based on that thought. In summary, then, the nature of the *Good*, or, more precisely, the nature of human psychology in relationship to the good, is what enables Murdoch's conception of the moral life as one that is never-ending. Despite the metaphors of sight, Murdoch's is not a philosophy that says one can grasp the really real or, in Kantian terms, the real *as such*. Central to her understanding

---

[44] Widdows, *The Moral Vision of Iris Murdoch*, 65.

[45] Antonaccio, *Picturing the Human*, 139. In particular, she adopts ethicist William Schweiker's *reflexive realism* to describe Murdoch's philosophy: "The point of understanding Murdoch as a reflexive realist is that she understands 'reality' as existing not only outside us … but mediated through consciousness and moral vision. Far from equating realism with the empiricist assumptions of the scientific gaze, Murdoch makes it clear that realism is always keyed to a personal vision."

of perception is that we see the world through vision and evaluative capacities already conditioned culturally and socially. The *Good* is not something one fully comprehends, fully achieves, or achieves union with. It may call toward us, but our dispositions toward selfishness are never transcended.

We can always come closer, then, but we will not arrive. One's moral development is never complete. At least in MacIntyre's earlier formulation of the moral life as a *quest,* there is the sense inherent in that term that something is achieved.[46] When one finds the Holy Grail, one has it in their hands and no further search is necessary. With Murdoch, however, the work is never done. The *Good* that is our goal is in its very conception something beyond our grasp. Indeed, in a significant way, this is an un-Kantian notion of ethics, at least in the sense that an *ought* does not illuminate one's capacity for reason and ethical action. Instead, any ethical expectation stands in judgment of one's capacity and efforts. It illuminates at every moment how far each individual still needs to strive, while also illustrating the ultimate failure of any attempt at moral perfection. This, however, is not meant to demoralize the moral subject. It is meant to spur one on. And it is to help in answering a pull opposing that of the innate egocentricity of humans, what Murdoch refers to as *Good.*

What this *Good* will look like will differ from person to person. (It will also presumably differ from culture to culture, era to era, shaped with and by various cultures, histories, and identities, even though Murdoch does not seem interested in the challenge to universal claims made by cultures.) This is not because Murdoch is a relativist, however. Humans are on a moral journey of moral development, and as their imaginative and critical capacities change, so will their appreciation of what constitutes the *Good.* The *Good* is not something one possesses, reaches, or embodies, but rather is something one reaches toward. Due to our finite human nature, there is always more to learn, more to change, more to regret, and more for which to hope. Our understanding of ourselves and what it is to resonate with this *Good,* then, is always prone to change. Moral and intellectual growth (and emotional growth, for that matter) are open-ended, as our lives and capacities, and our potentials, are open-ended. This means that if we are to agree that Murdoch's idea of moral growth toward perfection is a pilgrimage, it must be seen as a pilgrimage whose endpoint changes to some

---

[46] This may be one reason why *quest* is dropped in his later work, which after *Dependent Rational Animals* emphasizes how we may lose whatever gains we may get through such a quest.

degree over time. We are, in other words, continuously working on what goodness is and how we should be in the world as our experience, personality, and subjectivity change.[47] Our concept of the *Good* will continuously change, even as we reach toward that *Good*. If it is a pilgrimage, then, it is one toward a goal that, due to our finitude, will never be fully achieved nor fully understood.

Murdoch's point, however, is not so much to argue for the possibility of moral perfection as it is to include the motivating ideal of perfection within an understanding of moral subjectivity that can also accommodate human egotism and our obvious imperfections. And this understanding of the *Good* resonates with the way we have been discussing the moral life as Murdoch understands it. The moral life is constituted by a daily transformation that occurs moment to moment in everyday relationships and that is cumulative. We learn through this process—if all goes more or less well over the span of our lives—what being good consists of and how to be a person that searches for such goodness. It is a moral phenomenology of tension, where one's consciousness—this is Murdoch's preferred frame—is the suspension between the possibility of loving attention and a journey toward the *Good*, on the one hand, and the constant pull of egocentricity, on the other. Our consciousness is altered through the alteration of sight and attention, as well as through notions of the human as imperfectible and where, no matter the cultivation, we are never completely freed of our selfish propensities, which will always pull against our aspirations toward the bright sun of the "*Good*." There is in Murdoch's metaphysical grounding of morality the necessity of moral life as a search, and a pull toward the *Good*, certainly, but also a struggle, as well as a clarification and intensification, requiring effort that is largely constant throughout one's life.

Vulnerability and Responsibility

Murdoch, then, emphasizes in her metaphysics a dynamic understanding of one's life as, potentially at least, one of constant moral development. This is important to the present study, which seeks to account for particular

---

[47] In one of his critiques of Maria Antonaccio's work, David Robjant argues that there is a difference for Murdoch between the *Good* and the *concept of the good*. I agree with this, as it gets at the distinction that Murdoch is making between the realist nature of moral knowledge and our ability to grasp it. We will have a *concept of the good* that will be varyingly accurate depending upon our virtuous capacities. At the same time, it is a concept, because we can never fully apprehend the *Good* as such (Robjant, "As a Buddhist Christian," 997–8).

experiences of profound moral change and extreme political violence. A conception of the self or character as relatively stable cannot account for experiences of moral change brought about quickly through radical violence. Instead, such testimony as I have attempted to illustrate challenges a more essentialist or even stable notion of character and selfhood. Murdoch's view of moral subjectivity, on the other hand, is highly dynamic. It sees moral subjectivity as a constant orientation and reorientation toward a perceived *Good*, where that perception itself is changing along with our orientation toward it.

Such dynamism is a precondition for any frame that is to make sense of morally injurious subjectivities. It must give an account of not only the ways in which persons proceed through life as a moral subject but also the ways in which such subjectivity is vulnerable. Murdoch's account allows one the chance to recognize and articulate several such vulnerabilities that are not just part of the world but even part of the structure of the moral life itself.

There is, first, an inherent cognitive vulnerability in the moral life, or at least an epistemological one, that is structurally central to the metaphysical fabric Murdoch weaves for us. This arises from human finitude and the human inability to ever fully grasp the *Good*, an inability that exists alongside the ability to nevertheless aspire to the *Good*. The aspiration of human moral knowledge is humbled by the ever-receding horizon of the *Good*, giving rise to an epistemological limitation and vulnerability. As Stephen Mulhall writes, "since every attained image of moral unity is haunted by a deeper or more truthful one, it must be regarded as provisional or illusory."[48] Comparing where we are to where we were, at times we can get a glimpse of how incomplete our appreciation of what goodness really was, even if we thought it was a significant moral achievement at the time. With the experience of moral growth, we can learn to expect that the moral achievements of every present can be surpassed in the future, just as they were in the past. There is always, then, a more perfect way to be, a more perfect moral vision, a more perfect grasp of the *Good*. This means that we will never see clearly, not entirely.

This limitation has a direct effect on one's actions and behavior in the world. Since we never see with perfect moral vision, we will always be acting morally in the world in a way that is insufficient to the complexity that faces us. We not only can never act without being fully virtuous and

---

[48] Mulhall, "Constructing a Hall of Reflection," 227.

morally aware, but we always react to certain events out of some degree of ignorance. We may see something that could be hopeful as a source of despair, or, thinking that we know everything, believe that an outcome is certain. This can create a false pride, if we think our success is inevitable. It can also create hopelessness, if we feel that there is no way out of a bad situation, or if something horrible is inevitable. Both are misperceptions based on a flawed epistemology and phenomenology—thinking we know when we cannot fully know and thinking we see when our moral sight is always ever partial. This can add vulnerability to the fact of our powerlessness over certain situations, adding poor evaluations to already difficult circumstances, events, and disasters.

There is a second, related vulnerability inherent in the structure of moral imagination: a constant vulnerability to fantasy. As we have seen, imagination and vision are central for Murdoch's theory of moral development. Both fantasy and its opposite are made possible through our being imaginative beings.[49] Murdoch's emphasis on imagination rather than will and action, then, is not just about correcting what she saw as an imbalance in the way that moral philosophy had been framed. Instead, Murdoch understands imagination as a central activity by which we either lean into fantasy or see through to a worldview more indicative of the *Good*. Although Murdoch is known for her focus on the metaphor of vision for philosophy, Antonaccio seems to be right when she writes, "it is imagination rather than vision strictly speaking which is the primary locus of moral transformation in Murdoch's ethics."[50]

Murdoch is quite clear that we find the world already saturated with value. We do not create our ethics nor our decisions in a vacuum (one of her critiques of existentialism). At the same time, "imagination and attention introduce value into the world which we confront."[51] There is a sense in Murdoch, then, that we act in a world already present with a history, and yet also construct a world through our vision and moral development, the quality of which depends on the degree of development of our moral faculties, particularly those of attention and imagination. Murdoch is not consistent on this, and she does not lay out a fully formulated phenomenology. If she

---

[49] Murdoch, *Metaphysics as a Guide to Morals*, 334–35.

[50] Antonaccio, *A Philosophy to Live By*, 106. I focus on imagination/perception, because this is central to Murdoch's thought, and also helpful to the present discussion. Murdoch did not, however, locate morality in one faculty, but instead, viewed the moral life as nearly synonymous with life itself, making it a more holistic notion.

[51] Murdoch, "The Darkness of Practical Reason," 201.

did, we would have a better idea exactly how she understands the concept of *world* and how it relates to her ethics of vision, attention, and imagination. She wants to affirm the prior existence of history and the world to avoid solipsism, and yet the emphasis of her metaphysics is on a philosophical anthropology based in subjective vision. *World*, then, seems to take on a double meaning in Murdoch's thought, as it does for *Good*, designating at times something more objective and at other times designating another name for our subjective worldview and ethos. This tension—or, at least, these poles—show how one can be part of the world in a way that is subjective but in which such perception and vision can always change. There are always possibilities for worldview beyond the horizon of our current subjectivity, which provides a metaphysic that allows for, even requires, moral and phenomenological development.

There is always, then, an opening for fantasy. Moral imagination, understood not as a separate faculty but rather as an emphasis that the act of imagining has a central moral dimension, is for Murdoch structured in a way that is skewed and always pushing one to fantasy, at least to some extent. More importantly, as we do not just look at the world but rather interpret it, imagine it. In this way, we are in a real way accountable for our understanding of the world and the behaviors and actions based on such understanding. We have, then, a responsibility for the world—the world understood here as the *world we see*—as the *world* is not just something out there but is a constantly changing vision affected by our own capacities, the imaginations of others, as well as the broader matrix of social institutions, structures, and what we usually call "natural" life and events.

There is a sense, then, that each of us is responsible for our own worlds, the one we see, our worldview and ethos that is envisioned through our imagination, perception, the cultural images and narratives we inherit, the social institutions and structures in which we live, and the reality that encompasses it all. We partake in our worldview, and even when conditions are caused wholly by other subjects and are beyond our control, we live in a world of our making, at least to some extent. In fact, it might be more precise to say that we are complicit in how we imagine the world. We are not alone in the creation of our moral subjectivity, yet we are still necessary and agentive in that creation.

Such a feeling of responsibility, itself, is also the ground of vulnerability, a vulnerability to feeling one has failed, that one has not done what one could. Such a sense of responsibility, which makes up much of our moral lives, can become a source of indictment on our identity, our character, our very being, when we err. We should have done more, we should have worked harder, we should have taken more time to care. If the commission or omission is severe enough, an individual can feel they have fallen so far from visions of how one should be and act, that it can turn into despair. It is a despair that one's irresponsibility was so severe that one can never recover it. One's character is too far gone or has been revealed as completely wanting. To put it in more Murdochian terms, that vision of the *Good*, which as a hallmark always urges us on to a better, more virtuous life, can with despair turn into a more severe judge of who we are, where we are beyond saving or mercy. Or, at the very least, this despair and indictment is what one can experience.

This gives rise to a Murdochian responsibilism of sorts. Indeed, even though *responsibility* is not a key part of Murdoch's lexicon, her writing as a whole pushes in this direction. One is, as she writes in one of her earlier pieces, responsible for the world one sees, even as she is clearly against an idealism that views the world as only a product of mind or imagination.[52] There is a sense of freedom here, "a renewed ability to perceive and express truth," as Murdoch puts it.[53] But it also makes us vulnerable to feeling keenly our perceived moral failures. One may even say that the more morally responsible one is, the more vulnerable one might be to despair.

Murdoch, however, also insists that the moral nature of our worlds is not of our making. We lack power not only over the way we see the world but also in the constitution of our very subjecthood, making us vulnerable in our very selves. This is the flip side of Murdoch's understanding of responsibility as a ground of vulnerability. We are also vulnerable to those factors that influence moral subjectivity and that are beyond our control to adequately influence or change. Let me turn back to the example of the Bosnian War to clarify this point. An example Ivana Maček gives is how fatalities became a part of one's daily life during wartime, and how this affected one morally. One informant spoke to her of how people slowly changed the way they looked at wartime deaths:

> In the very beginning, every person killed was reported in all of the mass media. As time passed—it may sound a bit cruel, but it really is so—we started getting used to all those victims, and people began to turn into mere

---

[52] Murdoch, *The Sovereignty of Good*, 37; "The Sublime and the Good," 215; "On 'God' and 'Good,'" 354.

[53] Murdoch, "Art is the Imitation of Nature," 256.

numbers. It was reported only so and so many killed, so and so many hurt
… And when we came to a stage when they would for example report: ten
hurt, and you would say: well, it isn't so many. Two or three killed—oh,
then it is not so many today. You know. But that is terrible.

Where once one inhabited a world where such events were much rarer,
and so more morally appalling, they gradually became routinized, normal-
ized, and less of an occasion for shock and moral outrage. In this new
world, deaths could actually be seen as a sign of a good day, if in small
enough numbers. That this would have been unacceptable in the light of
older norms was not lost on Maček's informant. Even how one views the
morality of life and death, which is seemingly so basic to one's self-regard
and identity as a "good" person, could change without one's awareness or
one's acquiescence. It is a reality yet remains "terrible."

The gradual tide of "numbness," as Maček describes it, shows how the
transformed circumstances of war changed the way that one viewed even
killing and murder. It marked a profound shift in one's worldview. It also
marked a shift in moral development in which a certain degree of murders
was viewed with relief or even happiness. Once unacceptable levels of murder in pre-war years were now the cause for hope and a degree of pleasure
during the war, as the entire context of social relationships and possibilities
was transformed. There is a change here in how one evaluates events and
even what and how one desires.

This is not to condemn those besieged in Sarajevo. It is, however, an
example of how extreme changes to one's context will create changes to
one's own moral subjectivity, revealing vulnerability to moral shock and
loss to be a constituent of selfhood. In this case, the informant not only
demonstrates the changing way that people viewed life and death but also
is very reflective about how she and others eventually accepted this change.

Maček's informant calls this change "horrible" and "cruel," and yet
even if the individual did little or nothing to cause such a situation, one
still must participate in the life of the war-torn city. The social relationships, the connection to others, demonstrate and are a part of our contribution to the world we perceive. In the case above, the individual's
contribution to that world is in perceiving a certain number of murders as
not just acceptable but good. This is "horrible" compared to former, pre-
war norms, and the informant suggests that she feels culpability in agreeing to the evaluation of the numbers of deaths.

Indeed, this is one of the subtle yet profound and enduring cruelties of violence found in sieges, concentration camps, and other curtailed environments during war. A person is forced through violence inflicted upon them to respond to the world in such a way that a certain degree of horror becomes acceptable, even desired. This is part of the monstrosity of such violence, that one can become, or feel one has become, complicit in the wrong one would have previously decried. The morally deforming character of such violent regimes has been attested to by many survivors of wars and genocides, in particular the Holocaust.[54] If moral development is seen as an education in desire, as it was in classical Greek thought, then the deployment of such violence to curtail social environments applies pressure to malform that education and turn one's desire toward previously undesired and undesirable objects. It is an anti-conversion, or at least, it can be experienced as such, and due to the extreme nature of the violence and the pressure it exerts on individuals and communities, one from which it becomes difficult to apostatize. Indeed, this language of conversion and apostasy will be seen in the following chapters as particularly relevant in the context of the Bosnian War, as well as tragic, as individuals were being forced into ethnic and religious identities through the use of violence.

There is, then, a felt sense of responsibility reflected in these words, which Murdoch helps us to name more clearly. Such responsibility is inherent in the structure of how we envision the world. The informant mentioned above did not participate in the murders nor did she encourage them. She did not want to be a person that desired news of death, no matter the degree or manner. She was, however, part of seeing a world wherein murder was viewed differently than in pre-war times, which in comparison to pre-war norms was "horrible" and cruel. The view was her own, shared with others, even if she neither willed it nor liked it. She did not will the war, nor did she will a change in norms and worldviews, but she participated imaginatively in seeing a different world. The informant uses, after all, the first-person plural pronoun to talk about the way norms shifted, implying agency. Even if conditions made it nearly impossible to resist such a new vision, it remains her capacities and faculties that executed the new worldview. She participates in the creation of new, unwanted desire, a self-betrayal of one's Eros.

---

[54] A good example is Primo Levi's idea of the *gray zone*, where he talks more broadly about the difficulty of contending ethically with all of those who, besides the devils and victims, live through and somehow are complicit with violence. Levi, *The Drowned and the Saved*.

This is the vulnerability of responsibility. One is never fully in control of that over which one feels responsible. Yet, responsibility is not something you can turn on or off. It is grounded in the very fundamental ways that you imagine yourself and the world. It is bound up with identity and cosmology. The vulnerability lies in feeling responsible for something one cannot change.[55] The same critical self-evaluation that allows one to strive for goodness and to respond to the pull of the *Good* continues. Maček's informant has changed, yet the old norms still persist, and so their indictment over one's inability to affect things one cannot affect are still felt keenly.[56] Although she did not want such changes and did not want to desire in such a way, such change occurred. She has become what she once would have seen as a less than virtuous person, perhaps a vicious one. And, as she participates in that world, and so sees herself as somewhat responsible for it and for her character, she feels doubly bad, doubly responsible, for a change in world and moral subjectivity over which she had no real power.

An outsider could insist, however, that individuals like Maček's informants are trapped in a terrible situation and that their responses reflect a world transformed by another's hands. At most, the situation is tragic, but surely responsibility for the violence lies elsewhere. To put it in a more Kantian formulation, if one truly cannot do something, it is unfair to say that they should. I would, as an ethicist, agree with this. We are, however, not investigating objective responsibility but rather the felt, subjective relationship to one's own experience. Despite objective protestations to the contrary, there are many who still feel responsible either for being a part of that erotic transformation, however unwilling, or for acquiescing by responding to the world in what they see as a less moral way. Yet, even in the midst of being changed, and without one's willing it, one still remains a subject. As already defined through Ortner's work, such subjectivity includes agency. This means that, as a subject, conditioned by outside dynamics and yet endowed with agency, one may feel responsible for the effects of such vulnerability realized through one's imagination and sight.

---

[55] This is taken up by Nancy Sherman in *Afterwar* where she sees moral injury in soldiers arising from a soldier's notion of their responsibility for a situation. Sherman argues that moral injury can arise when the individual takes too much responsibility for a situation. Responsibility implies for Sherman the ability, the power, to affect a situation to a certain degree. The morally injured often have less power than they think, and so therapy requires a reevaluation of their responsibility, their power. Limiting a sense of responsibility is seen as limiting moral injury's despair (see Chapter Four, "Recovering Lost Goodness," in *Afterwar*).

[56] The persistence of old norms in new situations is an insight taken from Veena Das, *Life and Words*.

Of course, what makes these new desires and worldviews so toxic is that one still remembers past subjectivities. Memory contains in it not only the dismembered present but also what seems as the more whole nature of what one remembers. The individual understands that norms have changed, yet still remembers and senses emotionally and bodily the pull of former standards. This is a tension between what is sensed as more objectively good and what the world now seems to require. It is a tension, however, that is not abstract but rather is located in the subject, the one that still cherishes former norms yet must live in a moral and social landscape altered radically by war. Such a feeling of responsibility is grounded in the very structure of moral subjectivity, vision, and imagination. That this is so helps give an account of why survivors of war can feel responsible for aspects of the war that observers would say is not their fault.

This is an account that goes beyond whether or not survivors choose to feel a certain way and beyond questions about the culpability of survivors during wartime. Instead, it emphasizes that the structure of an individual's development as a moral being can make surviving political violence itself a double-edged sword. One can survive but with a moral remainder that makes living as a survivor a heavy, and for some a dangerous, challenge. We can, however, go deeper into this discussion and probe what these tensions are and whether we can better articulate them. To do this, I turn in Chap. 4 to Murdoch's representation of moral experience as one of tension to further detail this feeling of having lost the ability to be good. Focusing on this aspect of Murdoch's thought will build on my discussions thus far and allow us to bring in the way that political events and social institutions affect the experience of being a moral subject. Bringing in notions such as *local moral worlds*, I will add more nuance to this discussion to better account for experiences of political violence. This will help to articulate more fully what happens to one who feels she has lost her ability to be good.

CHAPTER 4

Tensile Moral Subjects

Murdoch's explicit writing on the nature of the tension of the self is quite short, consisting mostly of a brief chapter at the end of *Metaphysics as a Guide to Morals*.[1] In this sizeable work, Murdoch builds up to these last few chapters, in which she wants to balance her optimistic picture of morality with one not only of tension but also of vulnerability. Indeed, the work seems to be designed to climax on the theme of tension.[2] The work ends on a firm yet chastened hope of goodness, yet the dramatic arc of *Metaphysics as a Guide to Morals* creates an important moment of doubt at its climax. By doing this, Murdoch places an internal critique within her thought that, although not definitive, attempts to safeguard her moral philosophy from being overly optimistic.

It is this chastening moment that is such a rich source for a virtue hermeneutic of moral subjectivity, one that can account for the experience of moral injury. Any such understanding needs to account for moral aspiration as well as moral harm. Since the present goal is to account and describe experiences articulated by Dizdarević and Maček, and since those experiences reflect in an important way a loss of normativity and even an entire

---

[1] At the same time, however, her major work, *Metaphysics as a Guide to Morals*, is very much concerned as a whole with tension, including that between empiricism (seeing things as they are) and metaphysics (the emphasis on an approachable yet unattainable Good) (Antonaccio, "The Virtues of Metaphysics," 166).

[2] Antonaccio, *A Philosophy to Live By*, 70; Mulhall, "Constructing a Hall of Reflection," 220–1.

© The Author(s) 2019
J. Wiinikka-Lydon, *Moral Injury and the Promise of Virtue*,
https://doi.org/10.1007/978-3-030-32934-1_4

world, such a hermeneutic of moral subjectivity must illustrate the ways in which aspiration can slide into an experience of harm—how a whole world can landslide into a midnight sea. Such a hermeneutic needs to represent subjectivity, then, in a way that illustrates the inherent vulnerability of being a moral subject in the world while also describing how one can change from someone who feels optimistic about their moral life and aspirations to one who feels such aspirations are beyond reach.

Such a basis for moral subjectivity can be found in Murdoch's vision of the self as a *field of tension*.[3] The push and pull between our disposition toward selfishness and a teleological turn toward the *Good*, already discussed in these pages, is not an abstract matter but rather is a living tension experienced in the negotiations of daily life. For example, it is easy to see, upon reflection, how complex the nature and number of the goods, loves, and hatreds there are in a human life. There is nuance, and what we see as "good" or even desirable is often plural and in competition, where one good is pursued at the expense of another. Indeed, one loyalty can be pursued at the expense of another, potentially alienating an entire community or aspect from one's moral life as a result. The stakes then can be high, yet the range of choice can be too narrow to climb to such heights and satisfy all persons, communities, and factors involved.

This is also a beginning to understand Murdoch's view of moral development, the processes, events, interactions, and engagements where moral subjectivity is constantly reformed. Moral development—the process of actually being a moral being—is characteristically tensile in structure. It arises in the push and pull between various places of account, desire, good, and loyalty that make up what I describe here as moral subjectivity. The individual learns not only how to discern and negotiate each moment as morally significant. She also learns how complex the moral life is, experiencing the tension between different loyalties that can sometimes have no easy, or readily available, happy resolution. As far as moral development, this occurs in the midst of day-to-day life. Larger, more dramatic moments and events are important, but this tensile structure is the structure of the everyday, the *local moral worlds* of Arthur and Joan Kleinman. One's constant development as a moral subject occurs as one is transformed through engagement with communities, ideals, and different worlds, always

---

[3] Murdoch, *Metaphysics as a Guide to Morals*, 492. Murdoch refers specifically, though briefly, to "a field of force, a field of tension, between modes of ethical being divided under the headings of axioms, duties, and Eros."

exerting effort to orient oneself toward the *Good* and trying to understand what that is, as much as is possible. It is a daily, mundane process, but in its aggregate, it shapes who we are and, as both Kleinman and Sayer would put it, what matters to us most.

At this point, I turn to the modalities that largely comprise Murdoch's understanding of moral selfhood. For Murdoch, these modalities represent the areas of account that make up the tension, as well as the loyalties we feel we have and that jockey for our love and attention. It is these modalities in tension, along with an understanding of moral development as made up of everyday, intersubjective interactions in local moral worlds, that will create a representation of moral subjectivity that is able to articulate and account for experiences of moral harm and change. At the end of the chapter, I return to examples from the Bosnian War, which will serve as a way not only to illustrate this approach, but also as an occasion to bring back in the anthropology of subjectivity and notions of the *everyday* and *local moral worlds* to demonstrate how this virtue hermeneutic can combine effectively with anthropological theory to articulate the moral dimension of experience.

Modes of Moral Subjectivity

As moral subjects in the world, then, our consciousnesses exist to a great extent as fields of tension, constantly negotiating tensions between issues of duty, desire, and hope, among others. Murdoch's moral subjectivity reflects these tensions as different modalities within our experience, modalities that represent different claims on our attention and that are not easily reduced to one another, modalities with distinct ways of being moral and relating to others. How she articulates these categories is to take the central categories of ethics—deontology, consequentialism, and virtue—and use them as elements to construct a model of subjectivity to account for the contested, even fractured experience of being a subject.

Instead of keeping with this traditional philosophical trinity, however, Murdoch opts for a quadrilateral to represent the internal diversity of the moral self. We register value and its loss, and the experience of such aggregation and loss, through this schema. Murdoch takes these ethical categories and builds on them, ultimately arguing for four modalities—*axioms, duties, Eros,* and *void*—corresponding roughly to utilitarianism, deontology, and virtue, with void being the additional mode. Each modality represents a different area of human life, loyalty, and relationship, each with its own goods, obligations, and visions of the human (both good and bad) that are largely irreducible to one another.[4] Murdoch does not, however, seem to intend to name faculties or some ontological characteristics of the mind or brain.[5] She argues for the unity of the self throughout her work against more fractured conceptions, yet she also stresses the difficulty of being a moral human being in the midst of a world of conflict and the temptations to put oneself egotistically at the center of it. These modalities then can be seen as reference points that reflect the way that moral subjects experience themselves acting in the world and changing as moral subjects.

Each modality represents the many ways of being a moral subject, and to express the moral dimension of certain experiences, which other modalities do not capture. The modalities complement each other, then, representing important, rather distinct elements of moral life, all of which are required. They pull at each other, and although never reaching permanent equilibrium, do form a representation of selfhood that reflects the tension and complexity of moral subjectivity. Each modality, then, must be kept in tension with the others so that no one modality dominates one's moral subjectivity. If not, extremism can result.[6] Take *axioms*, for example. Axioms are those aspects of moral consideration and action that are based on foundational political principles. This can include utilitarianism and principles, assertions about general social happiness.[7] They also include human rights,

---

[4] Murdoch, *Metaphysics as a Guide to Morals*, 492–7. This is my interpretation of what the implications are for an understanding of the moral self interpreted through these modalities. Murdoch has claimed that we inhabit images of the human that we have created. Placing that claim in conversation with her understanding of a tensile ethical being, I interpret this ethical being as moving toward those images. Yet, as there are so many areas and spheres of life, value, and action, I make these images plural to reflect the plurality of moral trajectories and experiences represented in these modalities.

[5] Antonaccio, *Picturing the Human*, 161.

[6] Mulhall, "Constructing a Hall of Reflection," 227.

[7] There is still a vagueness to the concept of axiom. Murdoch writes, "It might be said that utilitarianism, intimately connected with politics, is an axiomatic philosophy ... As an axiom the utilitarian idea could be expressed as: 'The question, how will this affect happiness, is *always* relevant to *every* moral decision." She also says that utilitarianism is in "close relation" though "not under the exact same heading" as axiom (Murdoch, *Metaphysics as a Guide to Morals*, 493). Murdoch also writes that justice and rights are part of axiom, yet later in the same paragraph equivocates, saying that justice would "roughly" fit within axiom. This may be because justice has classically been seen also as a virtue, yet is often used as a social teleology or horizon, or perhaps because of the many ways it is now used—including discourse focusing on social justice—and which could fit into any number of moral categories or modalities. This equivocation could also reflect the fact that, although Murdoch wants to

as well as some understandings of justice. Axioms are almost statements of faith, assumed principles that are not reducible to more foundational assertions. They are the background commitments from which discourses like that of human rights emerge. For example, one's foundational belief that each individual is sacred or must be treated with respect is articulated through human rights regimes that aim to secure the rights of individuals against states and other institutions.[8] Indeed, Murdoch seems to be nervous that the individual can become lost in larger systems and theories, even her own. Her own emphasis on morality being something akin to pilgrimage, even mysticism, runs the risk of eclipsing the individual through processes that makes people "decreated," as Murdoch quotes Simone Weil.[9] In politics more generally, Murdoch wants to make sure that great movements, whether it be Marxism, fascism, other forms of totalitarianism, or even too-earnest liberal parliamentary democracy, does not totally subsume the individual into higher goals. Axiom can act as a "battle flag, or a barrier," fighting for important principles or trying to block injustices.[10] In an important way, this concept of axiom ensures within Murdoch's schema that the human individual is accorded dignity in politics, a concern that Murdoch has with what she calls the Hegelian tendency, which stresses the totality above the individual, losing the individual in the process.[11]

create relatively clean "moral modes" that are "divided against themselves," such a clean division is not possible within the complexity of the moral life lived with others (Murdoch, *Metaphysics as a Guide to Morals*, 492). It may also be that Murdoch is struggling with the fact that such categories are hard, in practice, to maintain, and any detailed discussion inevitably leads to connections and overlap. At the very least, this means that any correlation between Murdoch's modalities and the different modes of ethics must be handled lightly and should not interpreted as strictly.

[8] Murdoch, *Metaphysics as a Guide to Morals*, 294, 350–1, 355–6, 360–1, 493.

[9] Murdoch, *Metaphysics as a Guide to Morals,* 354–5.

[10] Murdoch, *Metaphysics as a Guide to Morals,* 356.

[11] See Antonaccio, *Picturing the Human*, 159–60. Murdoch discusses axiom most fully when trying to differentiate between "politics" and "morality." Morality seems to be her conception of moral development, modeled on the idea of religious pilgrimage (Murdoch, *Metaphysics as a Guide to Morals*, 367). Politics has more to do with formal, state-level policy and legislative decisions. She does speak about politics in a different sense when she says that politics can be found in the family, and that morality can be found in politics (Murdoch, *Metaphysics as a Guide to Morals*, 355). Her concern seems to be, again, to use tension as a way to safeguard certain areas of value or life. Although she understands morality through a metaphysics influenced by Plato and even Christian mysticism, she also understands that morality is not only about perception and attention, but also requires principles, human rights discourse, and notions of duty, which cannot be reduced to the arena of virtue, say.

On the other hand, it is in the nature of fundamental axiomatic principles to be propositions so strongly felt as given that they can overshadow class, race, and other considerations.[12] Such axioms, even an interest in human rights, can always be too sweeping, reductive, and so dehumanizing, if they become totalizing to one's worldview and subjectivity. Meant to circumscribe the threat of totality in ethics, axioms can recapitulate this threat in another guise. If one's subjectivity is driven by human rights as axiomatic, for example, and pays little or no attention to the realities and history of race, the human rights that are achieved stand to leave behind marginalized persons whose indignities are determined largely by histories of racism. This, then, is an example of extremism in one of the modes, of tightening this cord too strongly at the expense of other considerations, other goods. If the axiomatic modality is slackened too much, however, we return to Murdoch's original concern, that one can lose sight of important political rights that affirm the dignity and worth of individuals.[13]

An axiom, then, refers to a range of political principles, including but not limited to utilitarian ethical shades. The modality of duty within moral experience describes what is in some ways self-explanatory. It also reflects a particular understanding of "duty" in philosophical discourse. Murdoch writes, "Duty then I take to be formal obligation, relating to occasions where it can be to some extent clarified … Duty may be easily performed without strain or reflection, but may also prompt the well-known experience of the frustration of desire together with a sense of necessity to act, wherein there is a proper place for the concept of will."[14] What is emphasized in this understanding of duty is the importance of the means as opposed to the ends in moral reflection, imagination, and action. Duty emphasizes the pull toward doing what is expected of you, even if that means that such expectations are put before the outcomes of whatever action one takes. It is also a pull against desire, an understanding of obligation as an imposition on one's actions that may pull against (even correct) what one might want to do otherwise.

Reduction in such a sense would collapse tension and deprive us of language we need to make certain claims, to advocate, and even to conceptualize and verbalize experience. In this way, she is fighting back against a type of epistemic injustice. See Fricker, *Epistemic Injustice*.

[12] Mulhall, "Constructing a Hall of Reflection," 222–3.

[13] This is similar to Sharon Welch's understanding of "cultured despair" where giving up on fighting the rights and need of others is a luxury for those who have supports and resources to fall back on. Welch, *A Feminist Ethic of Risk*, 103–22.

[14] Murdoch, *Metaphysics as a Guide to Morals*, 482.

Philosopher Stephen Mulhall has a helpful reading of the difference between *duty* and *axiom*. Murdoch defines duty here in a way that is, as Mulhall writes, less involved than a term like axiom, which includes a larger, complex political landscape. Mulhall describes duty as inhabiting more personal, everyday experiences, while axiom works more in a political realm. He writes of Murdoch's understanding of duty that "moral rules [duty] are personal, setting concrete tasks for individuals in everyday life; axioms are public banners flown for complex social reasons that may often be grossly pragmatic."[15] Murdoch limits duty because expanding duty conceptually could "blur" the lines between other modalities of morality.

We could push back on this understanding of duty and argue instead that political issues do very much concern the everyday. This would be correct, but I believe that Mulhall is right, however, in that Murdoch conceived of these two categories separately so that she could capture different levels and modes of moral subjectivity. Differing between duty and axiom categorically allows her to stress the "political" which is understood as one of bedrock principle that intersects more obviously with social dynamics. We could always say that this involves duty, but it would not be duty as personal obligation, as Murdoch understands it. In this way, Murdoch is able to acknowledge traditional ethical concepts such as duty and will, which are so central to philosophies that she is critiquing, while limiting their role within her thought.

The danger of duty is that, if too rigidly held, it can also overwhelm other considerations that are also important. For the person of duty, of honor, their duty can become everything. Although important, this leaves out obligations to other aspects of being a moral subject, including that contained in axioms. One's subjectivity, dominated by duty and formal obligations, can result in "following a rule without imagining that something more is required."[16] Relationships to others, as well as to other goods and institutions, become translated through this limited understanding of obligation and morality. This, too, can be dehumanizing, as the needs of particular people in particular situations can be subordinated to the principle being followed.[17]

---

[15] Mulhall, "Constructing a Hall of Reflection," 223.

[16] Murdoch, *Metaphysics as a Guide to Morals*, 494.

[17] Duty is also challenging because it covers so much conceptual ground. One can talk about duty in terms of the citizen and issues of politics, but we also speak of duty within the context of the family, between strangers, and in contexts that are not first and foremost politics, or even largely political. Murdoch acknowledges this slippage even as she advocates for

The third modality is *Eros*, a central concept to Murdoch's project.[18] When Murdoch speaks of virtue, she is speaking of the Erotic aspect of experience, the area she sees as constituting the lion's share of the moral life.[19] *Eros* and the Erotic here are not specifically sexual in nature, though this can be a part of it. Instead, *Eros* is the realm of desire broadly construed and the area of moral experience to which the virtues most readily apply. It is the area of our "appetites," our needs, desires, but also what we find repugnant, disdainful, and where a great deal of our daily energies is spent.[20] It is an emotional and aspirational stance that is a key driver of thought, action, and development, and when we speak of ambition, love, regret, guilt, shame, hope, despair, and desire, it is the Erotic modality we are speaking out of. Although *Eros* involves virtue, then, it really represents the broader context in which virtue, strictly speaking, is relevant. It is as broad as desire, and duty as an obligation can be seen potentially as the tension between what one wants versus what is required.

In each modality, a main concern of Murdoch's is absolutism in the moral life. This can often be seen with a concern with totality, that is, the obliteration of the needs of the actual human through a focus on systems, principles, and more abstract obligations. Even *Eros*, however, can be absolutist, if one shirks their duties or ignores important political requirements, such as human rights. Too much passion and desire, too narrow a focus on unmediated duty, a focus on the greatest happiness for the greatest number (without looking at those left out), can create an extremism in one's orientation to the world. One can start seeing the world in stark contrasts and without nuance, obscuring the complexities, nuances, and uncertainties that define ethical life. If any one modality becomes too strong, while the others slacken, this can lead to a worldview that is dis-

---

duty as part of a vocabulary that can make distinctions between different areas of human interaction and value. Axiom, on the other hand, she restricts "more narrowly" and that is "better thought of as separate from personal morals and not assimilable into that sphere" (Murdoch, *Metaphysics as a Guide to Morals*, 356).

[18] Plato is an inspiration for this modality. Murdoch spent much time wrestling with Plato, which can be seen in her early tome, *The Fire and the Sun*, in which she argues, in part, to make room for the arts and literature within a neo-Platonic framework.

[19] Murdoch, *Metaphysics as a Guide to Morals*, 497. The scope of *Eros* changes between *The Sovereignty of Good* and *Metaphysics as a Guide to Morals*. In the first, Murdoch pushed back against duty as definitive of morality and ethics. In the latter, she concedes that duty is important, as it covers aspects of the moral life that the Erotic, and so possibly virtue strictly speaking, does not.

[20] Murdoch, *Metaphysics as a Guide to Morals*, 497.

torted. Such a worldview will be constituted without taking into account the other moral modes, all of which are needed if one is to have accurate moral vision and wiser courses of action.

There is in this conception of moral subjectivity, then, a sense of domination, where one mode dominates others, which can translate into one's actions in the world. Indeed, there is a connection to be made, then, between one's moral being and domination as it takes place between others. Survivors of concentration camps during the Bosnian War relate how camp guards, who after all were overseeing their neighbors whose faces and even lives they knew well, could start out uncertain only to grow over time into cruel taskmasters. Whatever psychological dynamics are involved, there is also a movement here in one's moral abilities and the modalities that, in part, define one's subjectivity. A growing, cultivated desire for domination that such an environment can afford can overpower notions of duty to neighbor, for example, as well as any human rights axioms. This could even overpower other Erotic desires, such as love or care of neighbor. This, to a point where one survivor described one such guard as not his own person but a creation of his nationalist movement.[21]

If absolutism in the moral life, understood as a form of domination of one modality, is one worry for Murdoch, then on the other end is a concern with a subjectivity that collapses into *void*, the last modality of Murdoch's moral subjectivity. This is a term with its own history in European philosophy, but Murdoch seems to draw specifically from Simone Weil's theological account of void.[22] For Weil, void is, in part, that experience of separation from God, which is important to Weil's understanding of spiritual growth.[23] Murdoch uses void, in contrast, to describe the moral experience that comes from loss, bereavement, or having everything important taken away from one, for which one may never receive

---

[21] Lieblich and Boškailo, *Wounded I Am More Awake*, 38–9, 76.

[22] There were other influences, however. Her engagement with Jean-Paul Sartre, the subject of her earliest work, influenced her understanding of void, or at least, influenced her to look more deeply into the use of such a concept. For example, in one of her essays, Murdoch defines Weil's void against existential angst. One is a spiritual achievement, the other is forced on humans. If this is any indication, the understanding of void could have been influenced by *angst* in that Murdoch's use of void seems to take the affliction that Weil speaks of yet the imposition and existential confusion embodied in Sartre's term (Murdoch, "Knowing the Void," 158–9). For an argument that Murdoch is much more of an existentialist than she claims, see philosopher Richard Moran's writing on Murdoch, including Moran, "Iris Murdoch and Existentialism."

[23] Weil, *Waiting for God*, 42.

relief or recompense.[24] It is an experience in which we question whether there really is something we could call "good." It is profound meaninglessness or its all-too-present possibility.

The multiplicity of this schema is important because it helps describe the complex experience of being a moral subject, and in particular the experience of a subject whose life is centrally understood as one of moral development, and where this development is understood primarily through the metaphor of tension. This does not mean that one's daily life is always tense. Everyday experience includes not only the experience of conflicting loyalties and goods, but also a perceived lack of such conflicts (even, perhaps, when an ethicist might say the individual should feel a bit more conflicted). One can give up on their obligations or too strenuously advocate a particular position. My point in arguing for such an understanding of moral subjectivity, instead, is to show a way to picture what happens when one experiences such conflict, as well as what happens when there seems to be no tension and no conflict. My assumption is that every individual is always engaged in the world in a way that can be represented through this schema of a tensile subjectivity and various modalities. This interpretation of moral subjectivity, cast helpfully as a dynamic image of tension and multiple modalities, can show why one may feel conflicted in the moral life, as well as when one feels unconflicted, and yet an observer might argue that one should. For example, in situations of conflict, one can chart the areas of conflict that create the tension. Perhaps, as discussed, one's dedication to human rights conflicts with one's loyalties and love of their family. Acting in a way that focuses on family diminishes commitments to more abstract human rights that concern people one does not know. We can imagine this as tension between *axiom* and *Eros*, possibly even *duty*. One does not want to become unfeeling, and turn to an axiomatic absolutism, but one also does not want to become parochial, and back one's family against all else. Usually, some middle ground or arrangement one negotiates with oneself of others, though the threat of becoming too absolute in one's subjectivity, or feeling like the moral life is impossible, threatening a meaningless void, remains present. One could also diagnose, to stretch that term, a person's ethical being, showing that in particular situations or, more broadly, environments, one should feel more tension, but instead, is seeing the situation or environment only through one modality, unaware of the loyalties and goods of other modalities that have a claim on her.

---

[24] Murdoch, *Metaphysics as a Guide to Morals,* 498–9.

We are never not in the middle of developing as moral beings. Instead, each individual in relationship with others—perhaps aligned with others or even against others—is attempting to inhabit more fully whatever modality is called for in a given situation. This balancing can prove difficult, as it is not always clear how one should respond or in what way. For Murdoch, this is the place where the development of moral vision and imagination is key, as they provide us with the growth necessary to be attentive to what is called for ethically. This schema, then, is a way to enter Murdoch's thought and to see her other concepts as connected to her understanding of subjectivity. For her, subjectivity can accommodate any number of actions one takes or orientations one has to certain goods, showing how these moves affect other places of account in one's life.

ETHICAL MODALITIES AND POLITICAL VIOLENCE

How, though, can this be used to better understand a particular period of political violence? Murdoch helps us chart the inner landscape of moral experience, yet such landscapes are never shut off from the world. If one takes the arguments of moral subjectivity seriously, subjectivity is always also intersubjectivity, as one is always in and of society and the larger world. Another way of putting this is to say that what Murdoch is describing is the experience of being a subject within local moral worlds, as Kleinman would phrase it. Neither in Kleinman nor Murdoch is there a naïve, dualistic psychology of an outer world and an inner world, however, but instead an understanding of the self that allows for interiority even as it affirms that one's inner life is shot through with the political, social, and cultural. Murdoch writes, "The moral life is not intermittent, or specialized, it is not a peculiar separate area of our existence. It is into ourselves that we must look ... The proof that every little thing matters is to be found there. Life is made up of details."[25]

This is a more intimate, personal, subjective way of describing what Kleinman tries to articulate through the idea of *local moral worlds*. For Kleinman, humans live on the local level of social existence, yet this local level is not just geographic but is also generatively normative. What Kleinman means by *"local moral world"* is one's immediate community or network. This is a broader understanding than a MacIntyrean ideal of community, which is narrowly focused interpersonally and even geographically. Kleinman argues with a more capacious eye that these networks are inclusive of not only villages but also urban landscapes, family settings, workplaces, and potentially even cyberspatial networks.[26] This idea of the local is *moral*, then, because it is in these settings, the context of everyday experience, where meaning, care, and concern are created and contested between people and within institutions. One's local moral world is the very place of evaluation, care, commitment, and norms, as well as their creation, contestation, adoption, and elimination.[27] Each individual, however, is complex, and has many identities and loyalties that are not reducible to just one identity. This also necessitates that one acknowledges the multiple moral worlds in which one lives. This makes *the local* implicitly plural, an umbrella term for the many local worlds one lives in, whether these be the home, school, work, a religious house, or other places that are the context of moral development and action.

Kleinman complements Murdoch in several ways, then. His understanding of the local not only as a sociological but also a moral-phenomenological space includes economic relationships and other modalities of human thought, action, and organization, yet privileges the moral dimension of experience and sociality. This emphasis on the local reflects Murdoch's own understanding of moral development as happening in intimate, interpersonal settings. As we saw with M and D, this occurred between two people, and at one point even occurred within one woman's own interiority. This is important in speaking of political violence, for even though macro analyses of politics and economies are critical to understanding violence, those who have actually suffered moral harm from such violence have such experiences in particular places, under certain circumstances, between particular people, within specific institutions, and feel the interaction and conflict between different levels of social interaction. At the same time, Kleinman's is a more sociological, institutionally grounded way of talking about the material ground for the experience of being a moral subject in the world. The concept of the *local moral world*, then, can be thought of as the space in which the moral vision, development, perception, attention, and the search for the *Good* that Murdoch describes occurs, drawing in the broader contexts beyond the local, such as the national and geopolitical, while emphasizing the importance of everyday life and experience. This gives the understand-

---

[26] Kleinman, "Experience and Its Moral Modes," 358fn2; see also p. 359.
[27] Kleinman, "Moral Experience and Ethical Reflection," 71.

ing of the self as one of tension more defined social contexts to understand how moral development is fundamentally a process resulting from profound interrelationship with the world on various levels.

This is important, because Murdoch does not spend much time on the institutional aspects of existence, nor does she spend much time if any on discussions of social structure. True to her vocation as a novelist, her emphasis is on the interpersonal and interiority. Indeed, her emphasis is even beyond the interpersonal. Important examples, such as that of the kestrel or works of art, exhibit how important engagement with the non-human, and even human artifacts, can be to moral development. Kleinman also emphasizes the interpersonal but does so with an emphasis not so much in charting subjective experience in Murdochian detail but instead on the interaction between the local level of existence, where morality arises, and other levels, such as the regional or even geopolitical, and how those levels act upon each other. Kleinman's notion of local moral worlds accounts for the ways in which intersubjectivity and social structure condition one's moral subjectivity. At the same time, Murdoch's account of moral vision and moral development provides vocabulary and concepts that help describe the ways in which the self is changed within such worlds and how those changes are experienced from the subject's point of view. Together they can enrich the current interpretive frame of moral subjectivity by providing language to describe moral experience and the context in which it occurs and is conditioned.

Using a frame that combines a way to articulate moral experience while attending to the conditionality of that experience, this understanding of the self as a tensile moral subject whose moral life inhabits local moral worlds can be used to better analyze and describe the ways in which moral subjectivity is undermined by the violence of war. We can see this more clearly if we turn again to the case of the Bosnian War. The wars of Yugoslavia at the beginning of the 1990s, and which gave rise to the Bosnian War in 1992, marked a thorough social crisis in Bosnia-Hercegovina, a fight for new institutions, and the transformation of existing institutions.[28] The ruling Communist Party in Yugoslavia during the Cold War had a vision of unifying the different groups in the western Balkans under a single, Yugoslav identity. Communist Yugoslavia attempted to transcend nationalist identities, such as Serb or Croat, with the multinational identity of Yugoslav, translated as *South Slav*. This proj-

---

[28] Cuk, "Temptations of Transition," 85; Vrcan, "The Religious Factor," 108.

ect was never entirely successful, and throughout Yugoslavian history nationalist identities grew. Only five percent of Yugoslavians would claim the identity of *Yugoslav*, even toward the end of a united Yugoslavia, preferring instead to claim a national or ethnic identity. Within a decade of the death in 1980 of Marshal Josip Broz Tito, who had founded and ruled the country since the end of World War II, this multinational project collapsed, allowing nationalist organizations, institutions, and symbols freer rein to transform and gain strength.[29] This included a contestation of symbols, with ethno-nationalist parties putting forward religious symbols to support their own strategies and legitimacy.[30] Sociologist Rogers Brubaker argues that such nationalism was "remedial," trying to solve the social crisis through the assertion of national institutions, organizations, and identities. In the end, the wars that erupted, including that in Bosnia, were, to a certain extent, conflicts between differing institutions, those supporting "Yugoslavism" and those supporting independent national identities, as well as groups vying over the power, meaning, and resources that certain institutions provided, or that new ones could give.[31]

The religious and national identities involved in this conflict were complex and changed throughout the twentieth century. Officially, Yugoslavia had three levels of political identity prior to the war. The first was that of *nation*, which included several, constitutionally defined national groups such as Serbs, Croats, Slovenes, Macedonians, Montenegrins, and, in the early 1970s, Muslims. Muslims were the only nation not to have their own republic, and the only nation defined religiously. (The constitution was edited in 1971 to include Muslims as a nation, with Muslims understood in "an ethnic sense.") The second level were the nationalities, which "had legal warrant to their own language and cultural assertion," but did not have their own republics. These included Hungarians and Albanians. The third level consisted of ethnic minorities, including Russians, Jews, Roma, Romanians, and even those who identified generically as Yugoslav.[32]

The countries to emerge from the collapse of Yugoslavia were all identified with the nations, or the main *ethnicities,* as they have been widely referred to in the English-language press. To oversimplify, Croatia was

[29] Bringa, *Being Muslim the Bosnian Way,* 34. "The dissolution of communist party rule had meant the breaking of the taboo issue of nationalism and national loyalties."

[30] Besirevic-Regan, "The Ethnic Cleansing of Banja Luka," 50; Perica, *Balkan Idols,* 336; Sells, *The Bridge Betrayed,* 107.

[31] Brubaker, *Nationalism Reframed,* 79.

[32] Bringa, *Being Muslim the Bosnian Way,* 25–7.

majority Croat, Serbia was majority Serb, and Slovenia comprised largely of Slovenes.[33] These nations/ethnicities had been privileged in Yugoslavia in that they were institutionalized within the political structure of the country. They had, then, the greatest institutional resources to create new nations out of their constituent republics, with the notable exception of Bosnia. Bosnia-Hercegovina, out of all of the states that comprised Yugoslavia, was the only one without a national majority. It had a plurality of Muslims, followed closely by a large Serb minority, as well as a sizable Croat population, among others. In the fight during the 1990s over which states would succeed Yugoslavia and how they would be comprised and defined, identities founded on nationalism and religion became regnant.[34] This was problematic enough in the other states, such as Croatia, with sizable national minorities, but it was even harder in Bosnia-Hercegovina, situated as it was between two new countries, Croatia and Serbia, each with majorities tied to the national minorities within. This gave rise to irredentism, with Croatia and Serbia supporting their co-nationals in Bosnia-Hercegovina, a dynamic that threatened to rip the new country in two.[35]

There will always be much to discuss about the conflict on a broad level, yet perhaps one of the most profound consequences of the Bosnian War was the transforming of *local moral worlds*. Larger armies were involved, but much of the atrocities and betrayals were committed by one's neighbors and others that one knew. The actions targeted largely the fabric of such worlds, and so the moral subjectivities they supported. Such violence touched directly on the bonds between neighbors and even close friends, and so the integrity of the networks that Kleinman says is so important for moral subjectivity. Journalist Barbara Demick, for example, recounts a man who noticed many of his Bosnian-Serb colleagues leaving town. One such colleague said he was leaving to participate in a hunting class. That same colleague was later caught as a sniper, one of those who

---

[33] Bringa, *Being Muslim the Bosnian Way*, 10; Cohen, 46.

[34] The landscape of identity in Yugoslavia changed dramatically before and during the war. Bringa reports that, although older generations would usually identify with a religious identity (*najice*), younger generations started identifying ethnically (*narod*). Religion had an ethnic sense before the war, but part of the battle, as we will see, was to more radically link specific religious identities with ethnic ones (Bringa, *Being Muslim the Bosnian Way*, 27).

[35] Brubaker has made the strong case that, in discussing ethno-nationalism, it is important not to unwittingly support ethno-nationalist assumptions by reifying groups in a way that reflects their division of the world into discrete, unchanging ethnicities. See Brubaker, *Ethnicity Without Groups*.

terrorized the streets of Sarajevo for years.[36] Esad Boškailo, a doctor who survived the concentration camps of the war, writes about the killing of his uncle by former friends.[37] Boškailo also relates how one prisoner in the camp thought they would not stay long as one of his friends, a professor, now worked in the Croatian defense ministry. Boškailo seems to realize what the prisoner does not, that it was the very friend in the defense ministry that had turned his friend in.[38]

These examples show the disillusionment, sometimes deadly, experienced in the war, in which trusted neighbors would become informants or even members of militias that took one's home and even killed one's family. This distrust, however, was so pervasive that it could affect relationships even within families where ethnicity and religion did not play so much of a part. Maček writes of one young man's experience, which she calls "typical." He lived with his mother, his kidneys were diseased, and his mother was very distressed. His godmother visited them every day, yet when she gave her money away, she gave it to a "refugee girl," instead of to the young man. His testimony implies that the godmother thought the refugee, displaced from her home into the perilous environment of the nation's besieged capital, needed it more. In his disappointment, he said, "It was logical that if we were close, if we were in some sort of family relation through godmotherhood, that there should be some priority in terms of who you are going to help." The young man eventually got the money needed for his medical care from another relation, but the episode made him question how much he could count on family.[39] Others helped him, invigorating his faith in other family members and neighbors, but Maček's purpose in this story is in part to point out the way that the moral ground beneath one's feet shifted quickly and dramatically during the war, altering relationships, obligations, and moral expectations.

Although Maček focused on the young man, it may be more insightful for my purposes here to focus on the godmother. It is safe to assume, given the lean resources available during the war, that her funds were also limited. After all, she was not able to give her money both to her family and to charity. Looking at her experience, we can see a tension between helping family, which includes *Eros* (care and desire to help loved ones), as

---

[36] Demick, *Logavina Street*, 10–11.
[37] Lieblich and Boškailo, *Wounded I Am More Awake, 53.*
[38] Lieblich and Boškailo, *Wounded I Am More Awake*, 13–14.
[39] Maček, *Sarajevo Under Siege*, 87.

well as notions of duty (obligation to certain groups). There is also a strong axiomatic mode present, as she gave not to her own but to the most needy. This move implies that the godmother had a notion of abstract personhood in which the individual is understood not only in terms of group or kin but also as bearing a certain dignity as a human being, as such. This reflects an axiomatic modality, though her decision could also reflect an Erotic modality, if she had been moved by the refugee's plight.

We can expect that there was tension in that choice and awareness on the godmother's part that she was neglecting one aspect of her moral subjectivity in order to attend to another, let us say axiom, at the expense of *Eros*. At the same time, with so much need, and so little resources, she would not have been able to attend to all locations of accountability to which she felt responsible. The young man and his mother did eventually receive help, but it came from other relatives. This only served as a contrast that further highlighted for the young man his godmother's failure toward him, both in terms of duty and care.

This story is significant because it illuminates the various modalities that were at play and how the very real tension between them can, especially during extreme situations, lead to suffering. For the godmother, there seemed to be no good action. Someone would have to go without during a time when a lack of charity could be fatal. The godmother could side with family, but that might seem selfish, especially if she saw that the refugee truly had greater need than her own family members. Or she could give to refugees, a move that was noble and perhaps from an objective point of view more just, yet nevertheless also constituted a betrayal of those she knew and loved. The effect of the godmother's actions on the young man was to throw doubt on his worldview. Where once he thought he could count on family, he now doubted the relationships in his lives and whether he could trust anyone in such extreme situations. For both, their relationships were frayed, and the intersubjective constitution of their local moral worlds, in which one's ongoing subjectivity and sense of value are constructed, were frayed. This is only one moment, one decision, but it shows how quickly and powerfully one's world can change during political violence and gestures to the way such tensions can transform one's local moral world and worldview and contribute to suffering and political violence. In such an extreme situation, daily decisions could assume outsized importance, as everyday tensions between places of account—say, between care of or duty to family and those to neighbors—could directly increase or decrease suffering or even directly affect issues of life and death.

This put extreme pressure on relationships and friendships, and, as Maček writes, friendships could change quickly during the war, changes that could last for years after the end of the conflict. Indeed, even acts of kindness to one person could be perceived as lack of care to another.

The son eventually received help from other family members, who, when giving him aid, also scolded him for not asking sooner. The son said, "I just sat and listened, and when they were finished I started crying, because we'd known each other earlier and I knew they were good people, but you know, during war people change."[40] These examples help illustrate how pervasive was the war's social corrosiveness, cutting away at the cords that bound together people's local moral worlds. It included not only neighborly relations but also relationships even within one's home. Relationships of trust that were once assumed could no longer be taken for granted. The significance of this can be difficult to imagine without such experience or without sustained investigation into such issues.

The pressure of narrowed or impossible choice, occurring so often, had an effect across the fabric of Sarajevo. It concerned a pressing question for many during the conflict, that is, whether one left or stayed in the city during the war and siege. Leaving was not necessarily an option for everyone, and opportunities could come or go depending on the progress of the war. As Maček writes, however, the choice to stay or go was "judged by strict moral standards. In the minds of those who stayed, leaving the city seemed like desertion or betrayal …"[41] In such situations, those who stayed disparaged those who left not for "their cowardice, for choosing safer lives and material conditions, or even for betraying the country and its ideology." Instead, they were "condemned for having betrayed the social codes of friendship, the city, its citizens, and the urban life they shared."[42] I heard this during my own time in Sarajevo, several years after the conflict, when one woman told me that she had no time for those who had left the city. It was their city, and if they left during the siege, leaving it to the invaders, they did not deserve to be there.

With the benefit of years and geography, both at a remove from the conflict, these judgments might seem severe or even cruel. How can one really judge another in such a situation? Indictments such as these threaten to cast a pall on any refugee, a consequence not intended by those who

---

[40] Besirevic-Regan, "The Ethnic Cleansing of Banja Luka," 87.
[41] Maček, *Sarajevo Under Siege*, 86.
[42] Maček, *Sarajevo Under Siege*, 116.

resented those who left. Indeed, the emotions and conflicting loyalties of the war were dynamic and complex, and it does no service to simplify the war's moral landscape. The condemnation recorded by Maček reflects a concern with an entire world and its passing. Was Sarajevo—a city recast during the war as a multinational, multireligious ideal—worth sacrificing for? Were the relationships of that world, defined in terms of tolerance, multiculturalism, and social generosity worth fighting for? *To leave,* understood in this light, becomes a deeply moral action, and even a practice, as *leaving* and how one *leaves* over the course of the long siege began to reflect how one valued one's neighbors and their shared world. This is more than just a question of patriotism, which, in the context of a war such as that in Bosnia where nationalist ideologies ran rampant, could be seen itself to be an exclusionary ideology. It was, instead, an unarticulated understanding that, by leaving during the war, one was abetting what the soldiers and snipers besieging the city were trying to do: destroy the world in which a multi-ethnic, multireligious Sarajevo was intelligible. It was a betrayal not of one's country, though that could also be true. Instead, it was a betrayal of a moral world that made the relationships and practices within Sarajevo intelligible and even possible.

This is a type of harm that philosopher Jonathan Lear has described in Kantian terms as destroying the conditions of the possibility of inhabiting a cultural identity. Looking at the experience of the Crow nation, Lear argues that there is a form of loss and of harm distinct from, yet connected to, extermination or even genocide. One can experience genocide yet still be able to practice their identity in another land. There is a form of resistance, for example, in survivors of the Holocaust being able to carry on their culture, even if it is in another land. What Lear argues is that in some instances the conditions that make cultural practices intelligible themselves are undermined. His example is of Crow warriors who, after the defeat of the Plains Indians at the hands of the United States, still try to count coups, horse raids, and respond to perceived slights from out of a warrior ethic. The domination of those cultures by the United States was so complete, Lear argues, that those practices, once seen as the very embodiment of honor, had become forever dishonorable. To take another's horse was now seen as theft. To attack a reservation agent was criminal. No longer was it possible to strive toward the ideals that are symbolized in the figure of the Crow warrior. Lear's argument is not meant to disparage the suffering of other cultures that experience genocide, but rather to bring out even further the depths of harm that violence can bring to a people. It is a form of harm that Clifford Geertz described, in which identities are made to be no longer "intellectually reasonable."[43]

What Lear does not discuss is the fact that not only were the Crow able to find a new subjectivity as Crow, but also that they were forced into the identity of "Native American." They were lumped together, through the dominant white understanding of culture, with all other Native nations, although their languages and customs were as diverse as those of any other region on the globe. One became not just Crow but Native American, just as people from Mexico and Argentina were "Latino," and all people from South, East, and Southeast Asia become Asian under the US racial imaginary. The Crow, Lear argues, found a way to reinterpret their symbols to "find a way where there was no way," a saying among African Americans to reflect the destruction of their subjectivities through the horrors of chattel slavery. But they were also forced into subjectivities that were not of their choosing, even though later they become very meaningful.[44] The ongoing struggle among "minority" groups in the United States, for example, attests to the challenge for dominated and marginalized groups to engage in identities ultimately shaped through domination.

This is the double-edged sword of survival for conquered peoples, as they often have to negotiate the identities that are put on them by the conquerors.[45] Indeed, the fact that Lear uses Aristotle to understand Plenty Coups and Crow history is itself symptomatic of cultural domination in the United States. Aristotle himself seems fine with slavery and domination, which makes him a problematic resource when speaking of dominated and conquered groups. Aristotle also helps make up the foundation of "Western" philosophy, and Lear never interrogates his own work to ask whether creating an authoritative account that interprets the meaning of a Native people's history which is itself a form of epistemic domination. In this way, he partakes of the long history of creating power over and by dominated groups by defining their history

---

[43] Geertz, *The Interpretation of Cultures*, 127.

[44] An example is the pan-Native American organizing of the American Indian Movement and more recently the pan-Native collaboration to protest the Dakota Access Oil Pipeline, whereby such identities can be used to challenge dominant cultural narratives.

[45] Lear also focuses on the work of one Crow leader, which can make it seem that Plenty Coups single-handedly made the interpretive leap that saved his nation. Seen as an example of how one can make such a move, it can be helpful, yet seen more literally as the heroism of one man, Lear's work can seem overly simplistic.

and making their experience and resistance an instrument for knowledge regimes foreign to those cultures.[46]

We can see Lear's concerns reflected in the case of the Bosnian War. In this context, the negation of ways of life began with the initial collapse of Yugoslavia. The identity *Yugoslav*, which transcended regionalism and nationalism in favor of multi-ethnicity, which no longer made sense without the supporting institutions and practices that gave that identity meaning. Without a Yugoslavia, there were no Yugoslavs. By fleeing, an individual—for example, a Bosnian-Serb—contributed to this important form of loss. If diversity was important to a multicultural Bosnia, one's absence decreased such diversity, and one's absence was felt by some, such as Maček's informants, as an implicit argument that either that multicultural worldview was impossible or was not worth defending. *Leaving*, then, did not just have ramifications for those who left. By leaving, one actively, if not intentionally, weakened another's ability to stay Bosnian or Sarajevan, and to resist other forms of identity that were being forced on others violently throughout the conflict.

More will be said on this in Chaps. 5 and 6, but for now it is important to note that those who chose to remain in the city during the war could view those who left Sarajevo or Bosnia as having stepped over a line drawn in the sand, a transgression, and perhaps betrayal that was perceived as threatening not just individual lives but an entire way of being and its associated and cherished values. It was a strike against fundamental meaning that enables one to make sense of the world and one's place within it as a moral subject. As Dizdarević wrote in a passage that sums up what was at stake in the war, "[The war in Bosnia] is the grandest and saddest funeral that has taken place on this planet in decades. Not because of the number of victims, but for the enormity of a whole mountain of irredeemably buried ideals." An entire way of looking at the world, including the hope that particular horizons of the *Good* are possible, seemed to be negated. "Apparently the victory of evil continues on unabated," Dizdarević wrote, "the powerlessness of good, the triumph of chaos over order, the verification of defeat in the match between humanity and the bestial goes on."[47] Seen in this light, *leaving* was a practice that undermined other practices

---

[46] These are not my insights but those I have learned from intellectuals of color who have reflected on these issues. See, for example, Andrea Smith, "Heteropatriarchy and the Three Pillars of White Supremacy."

[47] Dizdarević, *Sarajevo*, 88.

that made possible an entire world. *Leaving* had an effect on the most central stakes in the ware for those who remained, stakes that Dizdarević articulates in profoundly moral terms. Care, loyalty, justice, and duty— these are terms that resonate with the discussion of *leaving* and what practices meant in the context of the war. Indeed, being able to live into horizons that are best described in terms of *duty* or *Eros* or *axiom* could be weakened by the practice of leaving, as such a practice ate away at the possibility of such horizons.

What these examples also show is how dehumanizing betrayal can be experienced in war beyond discrete examples of torture and rape, which Bernstein focuses on. Rape and torture are common practices in war, and in the conflicts in Yugoslavia, rape was used as a form of warfare to dilute another group's perceived ethnic purity. In a conflict defined in terms of ethnicity, childbirth and a woman's body assumed special and dreadful strategic significance. In raping a Muslim woman, for example, it was thought by some Serbian nationalists that any child resulting from the rape would be Serb and not Muslim, thus colonizing the future and hastening the elimination of what the Serb nationalist ideology perceived as a Muslim race.[48] As awful as such practices and ideologies are, the earlier example of the godmother and the son show that extreme political violence, such as a siege, can create an environment in which the conditions for the possibility of trust themselves are worn away. The son was not raped or tortured, as far as we know. The war and siege, however, limited choices in such a way as to make people choose between various irreducible goods. In a situation such as that of the son, this could be enough for him to question who he could trust, as well as his own worth to those who should have valued him above all others, and to start altering his moral subjectivity. On the godmother's side, this could make her wonder whether, if there are no good choices, could there possibly be a life lived toward goodness?

An important element of the strategies that drove the war, then, was to destroy not only individual lives or even the lives of groups but also the possibility of certain communal identities, at least within certain political geographies (such as Bosnia-Hercegovina). The attack on such possibility targeted the sinews, both affective and institutional, that made up the ties of community. There are, for example, many accounts of people raped in

---

[48] For discussion of sexual violence in the context of the war in Bosnia-Hercegovina, see Allen, *Rape Warfare*.

front of family members, daughters in front of fathers.[49] Targeted people were made to feel afraid and uncomfortable in public. For example, in Srebrenica before the infamous massacre there, pigs were painted with the initials of the Muslim nationalist party and left to roam through the streets as an affront and threat to Muslim residents. The targeted were also forced to submit to primitive conditions: they were deprived of electricity, food, and all means of providing for themselves, and were also used as slave labor.[50] Half of the population of Sarajevo fled the city during the war, and about 150,000 internally displaced persons fled to Sarajevo from other parts of the country.[51] They fled not only the violence but also areas that they and previous generations had always thought of as theirs and wherein they felt they belonged.

An important consequence of these acts was to taint the imaginary of home and land with blood and horror. Where once one's hometown could bring to mind images of friends, families, and networks, in addition to the difficulties of any locality called "home," these strategies could rewrite those impressions so that home could inhabit a place of terror in shared social imaginaries. It could make it harder to inhabit or return and re-inhabit such spaces, alienating one from the actual material geography that provided the basis for one's moral community. The strategy of *ethnic cleansing*, a term coined during the war, was not only to eliminate groups from a political geography but also to make it difficult for them ever to return. Strategies targeting the local moral worlds, then, could make it appear impossible that one could ever have a community that could support the moral life in such areas.

Indeed, if a local moral world is to have any intelligibility, there must be spaces in which such a world can exist and be maintained. Individuals need to meet and exchange to some degree, and the very fundamental sociality that conditions one's moral subjectivity was damaged during the violence. The inability to simply walk down the street and greet a neighbor during the war destroyed such basic institutions as handshaking and broke down social interaction and bonds. Human Rights Watch witnesses said they were so frightened that they would rarely venture out. This meant they would go for months without seeing a next-door neighbor and could no

---

[49] Besirevic-Regan, "The Ethnic Cleansing of Banja Luka," 108.
[50] Besirevic-Regan, "The Ethnic Cleansing of Banja Luka," 75–78.
[51] Maček, *Sarajevo Under Siege*, 85.

longer participate in public life.[52] This increased fear, as one did not know the daily events in the city, nor what happened to friends and family.

Samantha Power, a former journalist and human rights author turned UN ambassador, relates how, in the environs of Banja Luka, a curfew was put into place for the entire night starting at 4 pm. She shows how public life was severely curtailed, indeed, eliminated, for Muslim and Croat residents:

> Non-Serbs were forbidden to: meet in cafes, restaurants or other public places; bathe or swim in the rivers; hunt or fish; move to another town without authorization [which usually meant giving over everything you have before you left]; carry a weapon; drive or travel by car; gather in groups of more than three men; contact relatives from outside Celinac (all household visits must be reported); use means of communication other than the post office phone; wear uniforms; sell real estate or exchange homes without approval.[53]

One could get shot randomly for simply stepping out in front of your house, and families could not even bury a family member in a local cemetery. Instead, they would have to bury their loved ones in the backyard.[54]

In addition to this, the social infrastructure—those material objects and buildings where community happens and is remembered—were destroyed, further undermining the constituents of one's pre-war moral subjectivity. Michael Sells, a scholar of religion, quotes a Croat militiaman who articulated well the mindset of the cleansers: "'It is not enough to cleanse Mostar of the Muslims,' said a Croat militiaman as his unit worked to destroy the bridge; 'the relics must also be destroyed.'"[55] Mosques, which are at the center of Muslim life, were torn down and dynamited throughout the country. In Prijedor, the mosque was torn down, and the Muslim-majority Old Town was shelled by artillery, destroying the distinctive Ottoman-era architecture and streets where Muslims would carry out their lives in public.[56]

---

[52] Human Rights Watch, "War Crimes in Bosnia-Hercegovina," 10.

[53] Power, *The Problem from Hell,* 250.

[54] Human Rights Watch, "Bosnia-Hercegovina," 8.

[55] Sells, *The Bridge Betrayed,* 93.

[56] Although I have discussed primarily Muslim targets, as they were targeted at different times by both Croat and Serb nationalist forces, Croats and Serbs were also targets of ethnic cleansing.

Memory, too, was targeted. In place of destroyed Muslim-owned buildings, Orthodox crosses were resurrected in the memory of war victims, not only destroying the evidence of local Islamic culture and community but also replacing it with another community's victimization narrative.[57] Tombstones in Medjugorje, a major Catholic pilgrimage site, were destroyed to erase evidence of generations of Orthodox family life in the community. And the first object to be mortared during the siege of Sarajevo was not a military barracks nor even the main mosque, but rather the national library, containing priceless and irreplaceable records of Muslim existence in southeastern Europe.

All of this demonstrates the atmosphere created in which the institutions and relationships on which community was based disappeared. Individuals could survive, but they could become isolated, and their community lay in ruins. In certain areas, space once filled with greetings, commerce, free movement, religious ritual, laughter, and argument was replaced by new memories and associations including betrayal, high anxiety, insecurity, blood, torture, and irredeemable distance between neighbors. Indeed, the community was tortured, terrorized, and slowly squeezed until individuals and their families finally left. It was a slow dismemberment that homogenized—cultural and religiously—communities across Bosnia-Hercegovina. And, like torture, the scars ran deep and forever maimed individuals and what community remained to them.

This is enough of a discussion to gesture toward the fact that the many factors that help sustain one's local moral world, in which one's moral subjectivity can survive, were targeted, frayed, and even destroyed. With the collapse of communication and access to others, one's world became quite small and different from what it had been. For many, rumor was the only source of information.[58] As Ivana Maček writes of the massive population exchange in Sarajevo caused by residents leaving and refugees arriving in the city, "The massive turnover in the population brought by the war affected social relations in dramatic ... ways, as kinship ties were ruptured, strained, or reinforced ..."[59] Neighbors, new and old, began to distrust one another. And they began to live in the narrower religio-

[57] Wesselingh and Arnaud, *Raw Memory*, 20.

[58] Palsto, *Media Discourse and the Yugoslav Crisis*; Wesselingh and Arnaud, *Raw Memory*, 38–39; Besirevic-Regan, "The Ethnic Cleansing of Banja Luka," 72; Rieff, *Slaughterhouse*, 58.

[59] Maček, *Sarajevo Under Siege*, 86.

national identities—Catholic-Croat and Orthodox-Serb—of the national ideologies that eventually took over many of the new post-Yugoslav states. Such a simple thing as a rumor, then, when it replaces all other outlets for information can increase fear and, after a while, can have an effect on one's perceptions of others. One will even come to see close neighbors with suspicion.[60]

It is difficult to be moral when the constituents and conditions of established morality disappear. What is more, there was a shrinking of the possibilities of the moral life throughout Bosnia-Hercegovina during the war. Where one might have once embraced Sarajevo's tradition of tolerance and pluralism, one might soon find him/herself xenophobic or distrustful of former friends. Where one once valued a more democratic government, one's axiomatic foundations may have changed to something less open. These changes in one's context, in social institutions, and even in access to the routines of daily life are moral issues. In order to survive, one might not feel the need to commit crimes but one had to see the world wholly differently than before. In the place of civility, there was suspicion; in the place of tolerance, anger. Dizdarević laments this when he writes, "The worst of it is that we have learned to hate. We have become suspicious, and don't trust anyone. We no longer know how to hope. We've become cynical and scornful."[61] For some, this created a dissonance between the person they were before the war and the person they had become, as well as a dissonance between what their world once was and what it had become. The Bosnian poet Semezdin Mehmedinović writes, perhaps with both insight and irony, "How can this happen/Here, of all places, where we're so humane?"[62] This can lead to feeling that one had lost something profound and important about oneself. One could no longer pursue the moral life in a way one had before, because that world no longer existed. How, then, can one hope to be good? Where does that leave the individual?

VIOLENCE TRANSFORMING SELVES

I have argued above that the political violence of the war also made it difficult, and at times impossible, to live into the various modalities that comprise one's moral subjectivity. Attending to one important principle, as we

---

[60] For a discussion of the way that rumor can incite violence, see Tambiah, *Leveling Crowds*.
[61] Dizdarević, *Sarajevo*, 102.
[62] Mehmedinović, *Sarajevo Blues*, 193.

saw with the godmother, could mean alienating family. Such experiences can, over time, wear away at one's faith, if you will, in the possibility of being able to orient oneself to the *Good*. To invoke Plato's allegory, no matter how one twists and turns, one seems only to see shadow. This undermining of the possibility of the tension that makes up the moral life can be exacerbated and quickened with the destruction of the material, social, and cultural structures, spaces, and institutions that support and inform one's various communities and the goods that they supply. Their destruction can also create conditions in which one's moral ideals and horizons are replaced with formerly cruel and horrifying norms, as with those who viewed a good day as one that included a certain number of murders. Even though the individual did not want such a transformation of norms, being a moral subject can include the cruel irony of feeling responsible for worldviews created by violent conditions beyond one's control and reprehensible when compared to former moral worldviews.

Political violence, then, can transform one's local moral worlds, and one's larger society, into new worlds that only admit of certain subjectivities and ways to understand the self. As Maček's informant stated, one can believe and live into worldviews that are "horrific" and "cruel." The individual can come to see herself as monstrous if the contrast between her past and present selves are strong enough. What one must do to live, and what one has become (as well as what one's world has become), might have seemed monstrous to one's former self. This is a source of suffering but also an indication to one of how much they have changed. The monster is after all, as the Latin root *monere* infers, a *warning* to others not just about the danger it can wreak but also the danger we have in our ability to become like the monster.[63] *Monstrosity* is a highly moral, normative term, signaling the threat to self and others that can come from moral harm. It takes not a special evil but enough violence to transform ourselves into beings we do not recognize and that we would have formerly thought monstrous. And life can become truly horrific, even desperate, if one lives into those former evaluations and thinks of oneself in the present as having become a monster.

With human psychology such as it is, we cannot live very long thinking we are bad or evil. Much has been written on how people who seem so objectively cruel and the cause of great evil will see the world, even warp

---

[63] "Monster, N., Adv., and Adj." *OED Online*. Oxford University Press. Accessed February 16, 2015. http://www.oed.com.proxy.library.emory.edu/view/Entry/121738

the world, into one where they are righteous. The perceived loss of moral ability, then, caused by extreme violence will spur one to extreme actions to respond to what is an untenable situation: the feeling that one cannot be good or move toward the *Good*. This helps us to begin to understand the questions raised in Chap. 1 concerning the relationship of violence, the feeling of not being capable of good, and one's moral subjectivity. The idea of the cyclical nature of violence can perhaps be seen here in moral terms, as the need to orient oneself toward the *Good*, which is necessary for one's very identity and the intelligibility of life and the world, and is so central that its contestation and deprivation can drive violence, even as it is caused by violence. One can, then, become monstrous within actions driven by the (increasingly desperate) need to be good.

The transformation or elimination of the material and social structures that support relationality and intersubjectivity can also eliminate former moral subjectivities, although not their memory. One can continue to feel bound to former norms, even though the devastated society around one will not support them or make it too dangerous to embody certain values, such as love, selflessness, or generosity.[64] It might also strain one's resources, so where it was once easier to be generous to many, for example, one's selflessness may in the context of violence be materially restrained. Any of these situations can create a feeling of having lost something essential.

We can see, then, how Murdoch's concept of the tensile moral self, the importance to the moral subject to strive for the *Good* and the multiple horizons that comprise it, along with Kleinman's local moral worlds as the context of such moral development and being, can help us begin to better understand the ways in which violence can erode a sense of moral ability. Such ability itself is based on a broader array of institutions, symbols, narratives, communities, goods, and so on, that support certain subjectivities as possible while making others more difficult. Indeed, some experiences, as we have seen, can support the very feeling of being able to move toward the *Good,* while others, particularly those of extreme violence that can maim society and the context in which one's morality and identities are intelligible, can be detrimental to such feeling. In such situations, trying over and over to be good, only to neglect certain modalities, can leave one feeling that one has lost the ability to move toward goodness, or even that goodness is no longer existent. Perhaps, one might even come to feel that

---

[64] This is an insight I take from Veena Das's ethnography of women surviving the violence of the Partition of India (Das, *Life and Words*).

moral effort was always futile and that one's experience is proof that the world is not capable of virtue. Such changes may not even take much time. Instead, the severity of violence can transform one's world that such feelings of moral loss become precipitate where for others it might take a long period of extended moral challenge and failure. And this feeling that one has been changed, perhaps irrevocably, can lead to despair and a feeling that one can no longer be good. It is a moral subjectivity that is best described with the modality of *void*.

CHAPTER 5

The Domination of Void

I have yet to mention the importance of art and literature for Murdoch's understanding of moral development, yet their relevance should not come as a surprise. Murdoch's emphasis on imagination, as well as her career as an award-winning novelist, lend themselves well to the place of the arts in the moral life. Along with certain intellectual disciplines (*technē*), art—particularly, literature—is one of the main methods through which we can learn about the moral life and what it is to be good. As we have seen, experience and education for Murdoch reveals standards and differences between better and worse, and the arts, particularly literature, are the quintessential means through which one can grow as a moral being. After all, we experience through art education how and why some art is better than others. Such experience allows us an understanding of how hierarchy is a central structure of our life, and this understanding can be reflected in our moral development, where we can begin to see how some actions and ways of being are more correct and better than others. This is a philosophy that is, indeed, unashamedly hierarchical and based on notions of authority, as Murdoch herself attests. Value is inherently derived from comparison, and through this process, one can eventually learn not only between better and worse but also worst and best. This is an imaginative leap made possible by building on the comparative to surmise the speculative. "We recognize and

© The Author(s) 2019
J. Wiinikka-Lydon, *Moral Injury and the Promise of Virtue*,
https://doi.org/10.1007/978-3-030-32934-1_5

identify goodness and degrees of good," Murdoch writes, "and are thus able to have the idea of a greatest conceivable good."[1]

How, though, does Murdoch's highlighting of art, beauty, and the kestrels of the world fare when confronted with questions of horror, suffering, and terror? What if one's world is not filled with kestrels and art but bombed-out museums, libraries, and homes; the trees gone and burned for fuel; and the mountains sinister with hidden armaments? What if what one sees, for instance, calls us to defensive re-entrenchment, fear of the stranger, or knowledge of the world as a dangerous place, where loving attention should be given cautiously and frugally? Or, what if one's daily choices pit helping one's neighbor against helping one's own children? Generosity could begin to look like a virtue with too high a price tag attached. If this occurs with enough intensity or over a long enough period of time, one could start doubting the very possibility of trying to be good in such situations, where every choice harms someone, perhaps mortally. Desperate situations can give rise to heroism, but they can also, perhaps more often, elicit fear and the practice of turning away from neighbors, a form of anti-conversion from the moral life Murdoch hopes for.

These issues reflect the experience of many in Bosnia during the war in the 1990s, as well as countless others in countless other conflicts. A good example comes from one of Maček's informants, who described how lulls in the fighting could reveal landscapes that, although the wellspring for works, such as *Guernica* and other twentieth-century art, are in reality landscapes shorn of art but also of trees, bodies, and kestrels, in addition to the infrastructure of socially manufactured beauty. The shooting ceased, but as Maček's informant describes it, "the town was very ugly looking … All is so destroyed … Only the skeletons of the stores, so much garbage in the town. A lot of concrete, cement, glass, everything."[2] If we then place

---

[1] Murdoch, *Metaphysics as a Guide to Morals*, 395 (emphasis Murdoch's). Specifically, Murdoch is paraphrasing what she takes to be Gaunillo's response to Anselm's original formulation of the ontological proof. Although Murdoch's emphasis on education and art education is important, it does not necessarily follow that Murdoch sees formal education as irreplaceable for moral development. The example of the kestrel is an important, democratic affirmation, if you will, in Murdoch's writing. Her emphasis on formal art and literature, however, raises the question of whether or not goodness is truly open to all, and raises the specter of elitism, at least in application. Murdoch, with her lack of systematic thought, leaves this question open (Murdoch, *Metaphysics as a Guide to Morals*, 175; Murdoch, *The Sovereignty of Good*, 54–5).

[2] Maček, *Sarajevo Under Siege*, 40.

the informant's account next to an understanding of the world as revelatory of the *Good*, such a juxtaposition prompts some sobering questions. They might include whether or not such experiences of violence may themselves negate Murdoch's attempt to see the world as good or her insistence on something we could call *Good*. Such questions have been central to philosophy and ethics over the past century, particularly after World War I and, again, after the Holocaust. A century that gave rise to the term "genocide" requires ethical responses that are on the lookout for unearned optimism.

Situations like that of Bosnia-Hercegovina during the war, and in too many other places to count throughout the last century (and beyond), are difficult to plumb because the experience of being a moral subject is both complex and interwoven with interiority, as well as the various levels of social institutions and interactions. Murdoch's notion of a void, combined with the larger understanding of moral subjectivity and moral injury developed so far, can provide an account of vulnerability which is able to help articulate how such moral loss occurs and how it feels. Thus far, I have focused on the other modalities of moral subjectivity to discuss the experience of violence, yet void is particularly significant for understanding the felt loss of self under conditions of violence. It is a category that can be used to capture the experience of extreme loss within the context of political violence, and when understood as part of the tensile moral self, it can also help represent the dynamism of selfhood, wherein such experiences can lead to moral injury.

Murdoch's understanding of virtue, self, and world has, then, not only an account of how one should proceed in their moral development but also a way of understanding, as well as articulating, times when such a process is derailed. Giving a moral theory that prizes image and metaphor is not, then, necessarily undone when images and metaphors are destroyed or malformed, not, at least, when such a theory takes seriously failure, fate, and fear in the moral life. Just as Murdoch accounts for how a kestrel can help us along, she also gives us a way to understand what can happen when those kestrels go missing. By showing the pillars of moral development, she helps to show where one's moral development is vulnerable. Insisting, for example, that the ethical life is one that is pitted against an ingrained egocentric nature; thus, we can see how a context of extreme violence may make the feeling of goodness a rarer achievement. If the only images available to our perceptions are so bleak and terrifying, they may seem to demand from us not the decentered selfhood that might come from birdsong on a blessedly balmy day. Rather, the desperation of such scenes may make a moral subjectivity of loving, generous attention the source of great, immediate vulnerability.

I do not want to deny their importance, but projects that seek to discern what an experience—such as that of Maček's informants, as well as that of Dizdarević and others—has to say ontologically about the world is outside the scope of the present work. I am not attempting to develop a type of existential or social theodicy nor create apologetics for a Murdochian metaphysics that can withstand the brute facts of violence and cruelty. I do think that Murdoch's philosophy has the bones for such work, but the task at hand is to show how a virtue hermeneutic can help us understand not the world but rather particular experiences of the world. My aim is to demonstrate and even illustrate what we can learn about such injury as a felt experience of one's own moral ability by supposing Murdoch's metaphysics and moral subjectivity. It is an attempt to apply a language and hermeneutical lens to situations of despair that shows the experience of such failings to be largely institutional and contextual, as opposed to personal and characterological—the latter being too often a methodological basis for judgment and scorn rather than understanding and compassionate insight. It is, then, a practical endeavor to help give an account of an experience of violence that seems to exceed the limits of current methods and discourses, both colloquial and academic.

Violence and Void

Extreme political violence as experienced in wars, such as the one in Bosnia-Hercegovina in the 1990s, is not a subject that Murdoch broaches explicitly. Murdoch did, however, witness firsthand the aftermath of some of the worst fighting of the twentieth century, and most likely had this in mind when writing about void.[3] Nevertheless, void as a concept accomplishes different goals within Murdoch's thought, and it is important to discern these understandings while creating an understanding of void that can reflect the experience of political violence.[4]

---

[3] Murdoch volunteered with the proto-United Nations organization, United Nations Relief and Rehabilitation Administration (UNRRA), which helped in post-World War II reconstruction, in Belgium and Austria after the war. Murdoch, *Existentialists and Mystics*, xix.

[4] Antonaccio, *A Philosophy to Live By*, 187. Antonaccio sees the discussion of the self as a field of tension as a summary of the *Metaphysics as a Guide to Morals*, and as *Metaphysics* is the

There is, indeed, a range of meaning of void in Murdoch's work, even though she discusses the concept only briefly in a few pages toward the end of her work, *Metaphysics as a Guide to Morals*. A void is, to speak generally, the experience of the possibility of meaninglessness.[5] It is an experience of the fact that we are not ultimately in control of our lives, even that our lives are defined in the final analysis not just by our birth and existence but also by death and the challenge that it makes to our pursuits, which may all be vanity.[6] It is an experience of our subjection to chance. "There is nothing that cannot be broken or taken from us," Murdoch states. "Ultimately we are nothing."[7] Indeed, void is "a counterpoise to happiness itself."[8]

As if this were not extreme enough, Murdoch goes further to indicate that such an experience can be nearly nihilistic. Void, in this sense, can articulate a loss of one's very self. Personality can be annihilated, or at least, one can feel like their personality—moods, dispositions, love, likes, dislikes, and hates—can be annihilated. Identity can be taken away. Deep bereavement or depression can result in the loss of one's interests and passions. What gave one meaning and joy are now meaningless. One can lose energy and stop acting as they normally would.[9] Something is taken—a loved one, one's personality, an ideal or ideal image, one's God—something that was critical to the existential ground that gives purpose to life, what Charles Taylor would call "the spiritual aspect of one's life."[10] Void,

---

last of her philosophical works, it also stands as a summary of Murdoch's final word on moral philosophy. Stephen Mulhall sees in it a prompt to continue thinking about Murdoch's thought, a place of continued profit (Mulhall, "All the World Must Be 'Religious,'" 34; "Constructing a Hall of Reflection," 239).

[5] Laverty, *Iris Murdoch's Ethics*, 97; Mulhall, "Constructing a Hall of Reflection," 238–9.

[6] Hall, "Limits of the Story," 9–10.

[7] Murdoch, *Metaphysics as a Guide to Morals*, 501.

[8] Murdoch, *Metaphysics as a Guide to Morals*, 498. Again, such a statement does not necessarily mean that Murdoch believes "we are nothing." At the very least, however, Murdoch is raising it up as a possible experience other can and could have and that that we must take seriously when discussing moral metaphysics.

[9] Murdoch, *Metaphysics as a Guide to Morals*, 501.

[10] Taylor, *Sources of the Self*, 4–5. Taylor's understanding of morality as a category of life and action is narrower than Murdoch's. Murdoch sees morality centered on an individual drama between one's selfishness and her striving for a higher attention to reality as it is and a loving attention to others. Taylor restricts this term to more traditional understandings in philosophy, such as one's obligations to other people, what we could see as duty (Taylor, *Sources of the Self*, 14). He reserves the term "*spirituality*" to refer to "what makes life worth

in this understanding, threatens not only the ground of the moral life but even that of one's life project in general. Indeed, "affliction," which is Murdoch's term for the experiences of void, she takes from Simone Weil's *malheur*, is a suffering involving loss and humiliation that is beyond daily sorrow or discontent.[11] It is a term that by definition separates more ordinary pain from extreme forms that can make one want to seek vengeance or to delude oneself that the lost loved one is somehow still present.[12]

Void, however, is also the base of an account of violence for Murdoch. Specifically, cycles of vengeance and violence are wrapped up in the wrong approach to void. Fantasy remains the arch vice or flaw (or at least is the result of an arch vice we might call self-indulgence that favors an egocentric worldview). Discussing void and Weil's notion of affliction, Murdoch writes, "We must experience the reality of pain, and not fill the void with fantasy. The image of balance: the void as the anguished experience of lack of balance."[13] The imperative against fantasy is that we may be so desperate to regain balance, to recover from humiliation or pain, that we may strike out in violence and anger. This is fantasy, where such visions of "bouncing back" can really restore one. This is why Murdoch eschews fantasy. Instead, she argues with Weil that we need to look at such pain squarely in the face and understand it in the context of the *Good*. We tend to ignore the Holy Saturday of our suffering and move too quickly to the

living." In this way, his understanding of spirituality and morality fit under Murdoch's much larger umbrella of morality that covers almost all of human experience (Murdoch, *The Sovereignty of Good*, 495). It is important to say that Taylor's philosophy is also more straightforwardly a teleology, whereas Murdoch does not have a teleology in the Aristotelean sense. If virtue is truly good for nothing, as she says, an understanding that other virtue ethicists, such as MacIntyre, agree with, it can become difficult to square this with the concomitant claim that the teleological *eudaimonia* requires virtue (Annas, "Virtue and Eudaimonism"). Murdoch turns to Platonic idealism, where we are never quite able to realize such ideals. Such a *telos*, or ends, then, motivates one to be good, yet also represents the fact that we will never be fully virtuous. Murdoch's framing of virtue is more about motivations than realizations.

[11] Murdoch, *Metaphysics as a Guide to Morals*, 502.

[12] Murdoch, *Metaphysics as a Guide to Morals*, 502. To an extent, then, *void* resonates with discussions in psychology about the death of the self under situations of extreme duress and dehumanization, as well as sociologist Orlando Patterson's understanding of "social death," where slaves are robbed of the power to decide their own life and daily decisions but also the meaning of one's life and actions (Gilligan, "Shame, Guilt, and Violence;" Patterson, *Slavery and Social Death*; Waller, *Becoming Evil*).

[13] Murdoch, *Metaphysics as a Guide to Morals*, 502–3.

resurrection. One can also seek vengeance, even "to hurt innocent people as we have been hurt," in an effort to feel better.[14]

There is for Murdoch, then, an ethic of suffering. The point of such an ethic is not to blame victims of violence for being a part of the continued cycle of violence. It is, instead, a reflection on the difficulties of the moral life and a warning to those who experience void not to appeal to cheap consolations that can too quickly rationalize vengeance. One must dwell with what has happened, seemingly because it is real and is actually what happened, and so the reality must be felt: "Instead of this surrender to natural necessity [fantasy, delusion] we must hold on to what has really happened and not cover it with imagining how we are to unhappen it. Void makes the loss a reality. Do not think about righting the balance, but live close to the painful reality and try to relate it to what is good."[15] This is for Murdoch, as ethicists Maria Antonaccio and David Robjant have both argued, an ascetic practice meant to use the suffering to transform our moral being[16]: "What is needed here, and is so difficult to achieve, is a new orientation of our desires, a re-education of our instinctive feelings."[17] Not all will be capable of this, however.[18] It is appropriate to call this approach an asceticism because it is a high call, as Murdoch seems to perceive. She holds out hope, even as she acknowledges the strong pull of cheap consolation, such as pretending that what happened did not or even providing an interpretation that does not honor the pain and stakes involved. Hope is not optimism, and it is not easily earned. It requires "re-education of our instinctive feelings," new desires. A saintly quest, indeed.[19]

I favor this reading. Murdoch has a dim view of basic human nature and most likely also appreciates the difficulties of life for so many. She did, after all, witness the devastating aftermath of World War II in Europe during her humanitarian work. This, however, is not what she wants to affirm. She wants to affirm, instead, an understanding both of how we can be good and how easy it can be to fail to be good.[20] Hers is a double

---

[14] Murdoch, *Metaphysics as a Guide to Morals*, 502.

[15] Murdoch, *Metaphysics as a Guide to Morals*, 503.

[16] Antonaccio, *A Philosophy to Live By*, 126.

[17] Murdoch, *Metaphysics as a Guide to Morals*, 503.

[18] David Robjant argues this, claiming that Mulhall and Antonaccio see Murdoch's understanding of goodness as having a magnetic pull on everyone (Robjant, 'How Wretched We are, How Wicked").

[19] Murdoch, *Metaphysics as a Guide to Morals*, 503.

[20] Blum, "Visual Metaphors," 322–3.

movement, creating yet another fruitful tension at the heart of her thought that reflects the tensile moral subjectivity she creates. Murdoch wants to account for a moral challenge and takes it seriously, but in the end, she wants to emphasize the ways through such challenges. After all, Murdoch not only saw the aftermath of war she also saw reconstruction and a Europe at peace.

I would, however, push Murdoch on the fact that she still wants to claim that we pass through most void experiences, they are short-lived, and can even be spiritually efficacious. Indeed, Murdoch does not necessarily universalize this, but she does want to affirm it. And, we can certainly see the truth in this; even after the loss of a loved one, time distances us from the fresh pain, and we can move on, to an extent. Quoting Paul Valéry at the very beginning of *Metaphysics as a Guide to Morals*, and later in the work when she discusses the ontological proof, Murdoch starts out the work with the claim that if difficulty is a light, then insurmountable difficulty is the entire sun.[21] The moral life is structurally difficult, and although she affirms extreme suffering, she also sees in it the fact that such periods in our lives can spur us on to wisdom.

In a way, then, there may be a nuance in Murdoch's thought between the mechanics of void and void as experienced. (I do not want to be too absolutist here, and so I wade into this discussion holding these concepts lightly.) Murdoch is saying that life is filled with suffering—even extreme, seemingly insurmountable suffering—but that there are ways through in most situations. This affirmation is distinct, however, if not fully separate, from experiences of the void that are extreme and that feel lethal to one's soul. That is, from one level void can be looked at within Murdoch's thought as an acknowledgment of the difficulty of the moral life but as part of a larger project that does, in the end, emphasize hope. On another level, that of the person dealing with experiences of political violence, this may seem a ludicrous assertion. The loss and suffering may seem too great, even for a long while after the events of violence have passed.

Murdoch seems to be right that, despite it all, most people seem to push through, and there is often something to be learned from the bruises life doles out. At the same time, void and despair can have the final say. Suicide would seem to be an evidence of this. There was a woman, Ferida Osmanovic, who escaped from Srebrenica, the only place to be legally labeled a site of genocide in Europe since World War II. Srebrenica was a United Nations (UN) "safe zone," where anyone within the city was sup-

---

[21] Murdoch, *Metaphysics as a Guide to Morals*, epigraph.

posed to be under the protection of the international community. Such protection failed, and units of the Bosnian-Serb army slaughtered the male population of the city. Ferida's husband was one of nearly 8000 men and boys massacred in and around the city in July 1995. The family had already fled from their town, which had been "cleansed" of Bosnian Muslims, arriving in Srebrenica as refugees heeding the rumors that the soon-to-be UN-protected enclave provided safety from the war.

Ferida's two children, Damir and Fatima, described the last time they saw their mother, the night after the men of the city were taken away by Bosnian-Serb militia:

> That night and for five days after, the air around Srebrenica was filled with the screams of men and boys being mutilated, slaughtered, some buried alive, others killed and dumped in mass graves; and of women and girls being raped. Damir and Fatima recall their mother becoming distraught. "At some point, she started repeating over and over again, 'My husband is coming, my husband is coming,' but perhaps she realized he was never coming back," Damir says. 'Then my mother said, "Stay there." We fell asleep and when we woke up the next morning we didn't see Mother around. My sister and I went looking for her. For two days we searched the camp, calling out her name. But we couldn't find her anywhere.'

Ferida's body was found the next day by another boy. She was wearing a white dress and a red cardigan, and she hung dead from a tree in the nearby woods.[22]

Concentration camps were also in use during the conflict on several sides. The experiences within could also strain hope to the breaking point. One example comes from a survivor of one such camp, Dr. Esad Boškailo, who relates his experience to the cowriter of his book, *Wounded I Am More Awake*,

> There were many ways to kill yourself in a camp. You could provoke a guard so he shot you. You could inflict physical harm on your own body.
> When Boškailo noticed a sudden change in a man in his early forties who had already lost his son and brother in the war, he started watching him more closely. First, the man stopped sleeping. Next, he stopped talking.

---

[22] Ferida's children would not know what happened for another 6 months, as officials did not know her identity. Lorna Martin, "Truth behind the Picture That Shocked the World," *The Guardian*, 17 April 2005. Accessed April 14, 2015. http://www.theguardian.com/world/2005/apr/17/warcrimes.lornamartin

> Then Boškailo saw the scars. The man was waking up each morning
> with bite marks on his arms. He was trying to bite himself to death
> at night.[23]

It is hard for me, someone who has never approached the severity of that
experience, to understand such desperation, and I read that passage won-
dering if there are emotions at play that I have never had to feel and for
which I have no name. The example speaks for itself as an example of
desperation and despair, but what is also interesting is that it brings
together the caveat I want to insist on for Murdoch, as well as her move
to hold out hope within experiences of devastation. In the case of Esad
Boškailo, who is a doctor, such experiences compelled him to form a
group within the camp so that the men could support each other through
the horrors. This helped Boškailo survive the camp, and he continued his
work with trauma survivors after the war had ended, which is itself a sign
of hope coming from such a grim time.

In a post-Holocaust world, it is unwise and uncaring to draw too much
meaning from violence. The above two examples, however, will always
stand out in my mind as a witness to utter despair. In situations where one
has lost everything, grace may be possible. But, life is such that it is too
much to say that grace will save all from the suffering of the world. There
is hope, but it does not always dominate all lives. Such examples carve out
a place in my understanding of void, where the failure of the intelligibility
of moral subjectivity and life itself is possible, even if we continue to hold
out hope. I want to affirm, then, Murdoch's project and use of void as a
way to affirm hope through an acknowledgment of suffering. For her, the
sun continues to shine; we will see it once more. Yet, even as Murdoch
holds out hope, the experience of void may be such that one cannot see
the sun. And though hope and goodness may yet be in reach of an individual, it may not seem that way. For this reason, it is important not only
to acknowledge Murdoch's claim but to also recognize that there are
experiences of void that indicate to the one experiencing it a lack of hope
and goodness in the world. One whose subjectivity is dominated by void
may not feel able to reorient themselves toward the *Good*, even if we philosophically argue that the ability persists. Such an experience of a morally
impaired self is real enough to make one's world, even one's life, unintelligible and desperate. It may be all-consuming.

---

[23] Lieblich and Boškailo, *Wounded I Am More Awake*, 43.

What I am moving toward is a reading of void that privileges more extreme experiences as its main constituents. This allows us to take the challenge that void was created to make—challenging the very possibility of meaning in one's life—while making it more coherent—focused on a narrower range of experiences whose connections can be more readily perceived. This will allow us to understand void in a way that, as we will now see, can help capture the experience of those who have experienced political violence and feel they have lost part of their moral being.

What Murdoch ultimately provides us with is, through balancing goodness and our propensity to egotism, a representation of moral subjectivity that engages these possible realities that we can still use in discussing political violence. Indeed, as such a representation is created in the middle of accounting for how the moral life can fail, it is particularly relevant to our study. As her representation is created to account for the void, she opens up a structure to emphasize experiences of void that can help us better understand the experience of political violence. Void, then, is a critical concept to describe the ways in which one can feel that one has lost hope and meaning through the activities of political violence. It gives a word to this experience, one that deals with the phenomenology of suffering and moral loss, while connecting it to a robust understanding of the self as a tensile moral subject always in the middle of projects that influence one's moral development and ability to maintain the intelligibility of the moral life and the efforts it entails. It helps us describe this experience and account for how and why one may feel they are no longer able to be "good."

ETHICAL TENSION AND DOMINATING VOID

One way to look at void is as a modality of moral subjectivity that can take over one's moral perception of the world through experiences of political violence. Political violence and related extreme suffering can collapse the tension of engagements with different modalities and moral worlds that people ordinarily negotiate. Such a modality brought on by violence can undermine one's ability to be oriented toward the *Good*, to take a term from Charles Taylor, so that a form of felt spiritual "atrophy" sets in.[24] One either no longer sees goodness as possible or it becomes nonexistent in some way. This is an extreme experience that affects one's moral subjec-

[24] Taylor, *Sources of the Self*, 107.

tivity profoundly, leaving one haunted by past norms in a world no longer able to accommodate their realization.

Let us return to Murdoch's representation of moral subjectivity, which includes four modalities, particularly void. Maria Antonaccio has argued that this fourfold schema is a summary of the aspects of Murdoch's *Metaphysics*, and we have noted that it corresponds roughly to common approaches in philosophical ethics to deontology, consequentialism, and virtue, as well as to the challenge to the intelligibility of moral effort embodied in void. We can, however, use this schema as a representation of moral subjectivity. The different modalities represent different ways of being moral in the world. Axiom, for example, represents the political dimension of experience. They also can be seen to correspond to different communities or thought traditions and ideals to which one has loyalty or is in some way obligated. Murdoch does not emphasize this, but these different modalities do imply different communities and traditions. If social justice as a good or value or moral horizon within one of the modalities is a central way through which one sees the world, then one is engaged in a tradition of thought and practice around social justice, and one is most likely part of, or sympathetic to, certain groups. This does not mean that one has to show up to meetings or be an organizer. We can think of community here more in Arthur Kleinman's understanding of networks as local moral worlds, as well as part of traditions. The ideals and visions of the human and society within these communities are internalized and influence the way that one sees political and social issues. If one goes against an axiomatic principle, one can feel real guilt or regret at having done so. Such emotions exhibit the ties between one's emotional and moral life and larger society.

Yet, these modalities, and the communities, loyalties, goods, and images of the human that they represent, can at times compete. This is fundamental to the tension that Murdoch finds in these different modalities. We are often pressed to choose between different goods in different areas of life, as well as between loyalties to different groups or ideals or obligations. Is it political action or family? Is it a feeling of having contributed to society or to love? Is it a new policy or is it living into the vision of a responsible family member one can depend on? Is it fulfilling a desire to see the world or a commitment to using your resources in a socially conscious way?

Such goods are not always in competition, but the moral life is necessarily one of tension and, importantly, of keeping this tension. If one were to devote one's all to being a parent and ignore both political issues and one's own needs, one could be charged with an unbalanced life. There are different ways, at least in the U.S. culture, to decry this, including saying that one's life is "one-dimensional," "out of balance," that one's priorities are "out of order," even that one is boring. It is a form of extreme life where one is devoted only to one area of existence, thus limiting one's experience and development as a moral being. Tension, then, as the effort to balance one's obligations to, and participation in, different aspects of life and different communities of concern is necessary to responsibility, in H. Richard Niebuhr's sense of the term: to be able to respond to the many places of account and obligation, as well as love, in one's world.[25]

However, void in my rendering represents not just an extreme imbalance of moral subjectivity. Void names an experience that raises the troubling possibility that the effort required to be moral and develop as a good person, one who is responsible to the many places of account in their moral world, is, in fact, meaningless. It is a threat of the negation of moral effort and moral development in a meaningful sense. Moments or situations will arise that will question the worthiness of moral development and the effort needed to constantly negotiate between different modalities and the communities and goods they embody. For most of us most of the time, we move through and away from such questions, even if some effects may linger. But void is a modality that remains and represents the challenges that being a moral subject engaged with the complexity of the world will place before our commitments, pursuits, loves, and hates.

More profoundly, void is a negation of identity and even personality. Connected to moral subjectivity is identity.[26] Each modality embodies not only goods but also images of the human, including ideal subjects, as well as images that are meant to warn or repel.[27] There are associated institutions, behaviors, symbols, and images that are contested and embodied. These make up our identity as political creatures, family members, dutiful people, and more. They reflect communities to which we belong and whose ideal and monstrous images of the human we live into or try to define ourselves against. For example, a central goal for certain belligerent

---

[25] Niebuhr, *The Responsible Self*, 202. This would include, according to Niebuhr, the need to expand and search out places of account that one may not know of. Responsibility, then, is not just being able to respond. It also involves a search for those locations, situations, or communities that one has a relationship of responsibility toward.

[26] Taylor, *Sources of the Self*, 28, 36.

[27] This also resonates with Murdoch's notion that we create pictures of ourselves that we then try to become. (Antonaccio, *Picturing the Human*, 45)

groups during the Bosnian War was to alter the moral and cultural landscape of the country and narrow the scope of people's identities. There was a genocidal ideology present in the competing nationalisms of the war.[28] Each pushed to enforce a dominant and narrow identity based on religion and ethnicity upon the population. The only identities that mattered were ethnic: Croat, Serb, and Bosniak (Muslim). And these were understood religiously. To be Croat was to be Catholic and to be Serb was to be Orthodox.

The issue, however, was that this narrow understanding of human subjectivity was not supported by the experience of most individuals. Everyday experience refuted it.[29] The importance an individual in Bosnia-Hercegovina accorded their ethnicity or religious heritage differed by geography.[30] But, everyone had multiple identities and loyalties. Depending on the context, we are family members, professionals, friends, neighbors, fans, and more. In other words, we all have a common experience of diversity, or at least of internal pluralism, as we each embody multiple, often conflicting loyalties, that change day to day, even hour to hour. Such experience can push back, and eventually will, against whatever worldviews or social hermeneutics we might adopt.

These issues are so fundamental to one's selfhood that these multiple identities, loyalties, and commitments cannot be negated without violent coercion. In Bosnia, such coercion was enacted at the price of human life, community, and any affirmations of the richness and robust, if tragic, pos-

---

[28] Sells, *The Bridge Betrayed*, 28, 51.

[29] For example, in his influential essays in *The Interpretation of Cultures*, Clifford Geertz draws on Max Weber's claim that experience can and will contradict the rationalized systems of meaning that communities create. At the heart of his theory of the rationalization of religion is the insight that experience can and will contradict the systems of meaning communities create to answer questions of inequality, deprivation, suffering, and domination (See Weber "The Social Psychology of the World Religions;" *The Sociology of Religion*; *The Protestant Ethic and the Spirit of Capitalism*). Geertz retains Weber's claim that experience can be in tension with the metaphysical and moral claims of religious leaders and doctrines. As Geertz states in his understanding of the problem of meaning, there is the possibility that every individual will experience moral inadequacy with our belief system, even with the world. (Geertz, *The Interpretation of Cultures*, 3–30) Whether or not one accepts Weber's broader claims, I want to affirm this idea that pluralism inevitably challenges the ability of our worldviews to provide an answer for every situation. It creates an aporia or type of moral or cognitive dissonance, which provides an opening that, although not always taken, can challenge our religious and ideological systems.

[30] Bringa, *Being Muslim the Bosnian Way*, 21.

sibilities of being alive. The targeted "other" or "out-groups" first had to be convinced of their identity as others. There were three main ways—elements of which we have seen previously—to enforce identity on the other and coerce them into abandoning their communities and homes:

> Radoslav Brdanin, president of the Crisis Staff of the "autonomous Serb region" of Banja Luka, which included Prijedor, in 1992 proposed "three stages of ridding the area of non-Serbs: 1. creating impossible conditions that would have the effect of encouraging them to leave of their own accord, involving pressure and terror tactics; 2 deportation and banishment; and 3 liquidating those remaining who would not fit into his concept for the region ... two percent was the upper tolerable limit on the presence of all non-Serbs in the region."[31]

Sociologist Rogers Brubaker corroborates this, stating that the ethnic cleansing and ethno-nationalist rhetoric "... involved the nullification of complex identities by the terrible categorical simplicity of ascribed nationality. It has involved essentialist, demonizing characterizations of the national 'other' ..."[32]

This strategy served two purposes. The first was to coerce members of an ethnic "out-group" to leave a specific area. The second was to make them suspicious of other groups and flee to their own.[33] The logic behind

[31] Wesselingh and Arnaud, *Raw Memory*, 35–6.

[32] Brubaker, *Nationalism Reframed*, 20.

[33] This also served to consolidate Serbian identity in Bosnia, as all groups had to be pressured to assume the religio-ethnic nationalist paradigm. To quote Noel Malcolm, a historian of Bosnia, "The main aims, clearly, were first to terrify the local Muslims into flight, and secondly to radicalize the local Serb population ..." (Malcolm, *Bosnia*, 236). In Prijedor, for example, Serbian nationalists faced an uphill battle to persuade the populace of its cause. In the 1990 elections, the Serb nationalist party received only 28 percent of the votes, even though 42 percent of the population identified as Serb (Wesselingh and Arnaud, *Raw Memory*, 38). Extreme efforts were needed, including violence. Malcolm has identified three methods used across Bosnia for the purpose of mobilizing one's own group on behalf of ethno-nationalism. The first was to "radicalize the Serb population with a non-stop bombardment of misinformation and fear-mongering through the media and the local politicians (Malcolm, *Bosnia*, 216–217). Early on in the conflict, rising nationalist leaders took control of media outlets, restricting alternative journalistic sources and feeding propaganda to their people" (Wesselingh and Arnaud, *Raw Memory*, 38–9; Besirevic-Regan, "The Ethnic Cleansing of Banja Luka," 72; Rieff, *Slaughterhouse*, 58). The second was the guerrilla technique of "compromising villages," involving "staging an incident—for example, shooting a carload of Croatian policemen outside a particular village—to invite a crackdown or reprisal, and distributing arms to the villagers, telling them that the police are planning to attack

this was that if you were attacked enough as a member of an ethnic group—for example, as a Muslim—you would soon run to Muslims for protection, even though you originally refused such an identity.[34] The Croatian writer Slavenka Drakulić described it as being "overcome by nationhood." She wrote that she came against her will to be defined by her nationality alone, confined, as she put it, in a shirt that cut off her blood and that did not fit.[35] Indeed, one of Barbara Demick's informants, reflecting on her experience, articulates not only the way in which violence can enforce certain identities and destroy others but also the way in which such changes continue through the generations. The informant told Demick, "We never knew we were Muslims before. The Serbs forced it on us, so now I try to remind my girls not to forget who they are."[36] Indeed, I would argue that Demick's informant is not quite right. There is a flattening of temporalities in her statement. She seems to equate who the girls are with whom they always had been, an identity that had been somehow hidden from them in the past. Instead, what she is really doing is reminding them of *what they have become*. "Muslim" is not so much what they were—an identity hidden and now revealed through the war—as it is an identity made urgently available to them through the war, and indeed, even one that was forced upon them. Violence was used to punish one for refusing or forgetting one's imposed identity and their membership in the political, ethnic group they had been forced on pain of death to identify with. It is part of the ideology of the war

them. When armed police do arrive, it is easy to spark off a gun battle; and suddenly a whole village, previously uncommitted, is now on the side of the insurgents." And the third was to create "violent incidents and then asking the army to intervene as an impartial arbiter, when it was perfectly clear that the army, with its loyalty to Belgrade and its Serb-dominated officer corps ..." (Malcolm, *Bosnia*, 216–17). In-group members were also threatened with retaliation if they helped members of the out-group (Sells, *The Bridge Betrayed,* 107). These were tactics that needed to use violence, threats and deceit in order to overcome the social bonds and institutions shared by different groups, even members of one's own ethnic group, who would not necessarily view their ethnicity as a primary identity marker, nor an ethno-nationalist worldview as a compelling interpretation of reality.

[34] "Certain dramatic events, in particular, can galvanize group feeling, and ratchet up pre-existing levels of groupness. This is why deliberate violence, undertaken as a strategy of provocation, often by a very small number of persons, can sometimes be an exceptionally effective strategy of group-making" (Brubaker, *Ethnicity Without Groups*, 14).

[35] Brubaker, *Nationalism Reframed*, 20. This is corroborated by data from "The Children in Crisis Survey" in Bosnia, which showed that in "ethnically mixed areas of Yugoslavia such as Croatia and Bosnia, religiosity was considered to have 'a certain compensatory and nationally defensive function,' especially when a small segment of one ethnic group found itself surrounded by a 'greater nation.'" (Cohen, "Bosnia's 'Tribal Gods,'" 52).

[36] Demick, *Logavina Street*, 71.

that makes such an identity, taken on because of the necessity harnessed through violence, seem like a revelation instead of a profound alteration to one's moral world and what is possible.

I include this example as it shows the ways in which political violence can transform identity as it is connected to one's moral subjectivity. The plurality of identity reflects the modalities of the tensile moral subjectivity. People have different identities, including political, cultural, familial, and so on, that reflect the modalities of Eros, axiom, and others. But, in Bosnia, a moral vision of the world was being imposed in which some identities and ways of being were approved, whereas others were deemed cause enough for violence, in which identity was being reduced to one dimension. The coercion targeted different modalities of being a moral subject, forcing one to narrow one's commitments and loyalties. Again, the elements of the local moral world can be damaged or destroyed. Such violence, then, is inherently and irreducibly of central moral import for the individual, in addition to whatever other aspects of life it affects.

The Corrosiveness of Political Violence to Moral Subjectivity

A subjectivity dominated by void raises doubts about the moral life that, in Murdoch's conception, is nearly synonymous with life itself. It is the place of desolation and despair, where one can lose all interest in what once made them passionate. One can lose their personality, so that close friends and family members will remark, "She is not herself." One can lose, then, their identity or aspects of it through moral crisis.[37] Self-loss, the loss of personality, one's responsibility (as ability-to-respond) toward those communities, represented in different modalities, that provide identity as political creature, as family member, and so on, can be the result of a moral subjectivity dominated by void. As it takes experience of sufficient intensity to throw into doubt that which matters most to one, we are thus talking about an extreme loss caused by extreme experience.

Returning to our previous question, what if one's surroundings and experience become populated with images of horror, of a society that is falling apart, where one sees not only the best of people—which happens during a crisis—but also the worst? What if that comes to dominate one's vision? As Stephen Mulhall reflects on Murdoch's field of force, "It is also worth noting that she offers orientation to her reader by means of an

[37] Taylor, *Sources of the Self,* 27.

image or figure, the idea of a field of force or tension; this not only suggests that the kind of unity she detects in moral experience is highly provisional or limited but also indicates that images are part of the tissue of her thinking—not an ornament or optional extra but the thing itself."[38] What we see, and the images that we value, are central to the makeup of who we are. And, in situations such as war, siege, and political violence, the ideal images of, on the one hand, a human being, and, on the other, images of the human as monstrous, which inform the horizons of one's modalities, can become reversed or be undermined. The good person may not seem a possible ideal to realize in situations in which there are not enough resources to go around. Alternatively, such ideals may still seem possible in one's eye, but the cost may be too high, and so we may come to see such ideals as a form of cosmic mockery or as a source of scorn directed at our moral inability. Goodness as conceived ideally in a pre-war world can come into conflict, for example, with providing for one's family and for caring for those closest to one's heart.

Such situations can create a tension that is impossible to live with. Do you let your children grow hungry, your wounded spouse go without needed medicine? Or do you steal? Do you even steal from those who are weaker than you, making your transgression of previously held values, such as charity, generosity, and responsibility toward the needy, all the more egregious? Do you not tell neighbors where they can find food that week so there is more for you and those you love? Perhaps such theft or lying leads, in part, to someone's death? The joy at feeding a family member may remain authentic, but it does not necessarily ease the pain of living into vice, if we can use that term, that one once thought as anathema, even inconceivable when applied to one's self-regard as an upright individual.

In such situations, it might seem impossible to be good. Violence has changed one's environment and with it the secure, stable conditions that enable one to live into their moral ideals and virtues more easily. The world is now a place where a good person cannot survive. It may seem impossible to respond to each modality, each community, each person or individual, or each good in a way that all deserve. Perhaps we begin to think that this is how the world really is, and any veneer of civilization or community is now revealed as a fraud. The effort required to be responsible begins to seem meaningless, as it is no longer possible to be responsible in the way we once believed. This is true in a quite literal way, as one might come to feel that it

---

[38] Mulhall, "Constructing a Hall of Reflection," 221–2.

is too high a price to respond to the needs of strangers and even neighbors. As we saw earlier with the example of the boy and the godmother, it may seem hard to respond even within the family bounds. This may continue until one feels that it is no longer possible to be responsible—to be able to respond to others. Such a situation could be figurative or even literal, as one could turn their back on others, cutting off response. We do things that we would never have done, and we begin to wonder who we have become. We begin to wonder if we will ever be able to be the good people we once were or even people who tried to be good and took morality seriously.

Of course, becoming someone new, someone forged in crisis, can be liberating for some people. Learning how to survive, overcoming the odds, can be exhilarating, as it can also reveal in one abilities one never thought they had. Newer, more valued subjectivities can arise. For others, however, it can be damning. And even those who feel exhilarated may be ashamed by it after the violence is over. Such situations are endlessly complex and conflicted. One can be as proud of their violations as they can be shameful of their joys.

Such experiences can either snap the tension of subjectivity or slacken it so that it is no longer meaningful or possible to try to be good. This can happen in a number of ways. In what we have been discussing, there are at least three different experiences of void we can discern and differentiate in the experience of political violence as seen in such cases as the Bosnian War.[39] These are: a feeling that one is no longer able to be good (that one has lost the ability, though it may exist for others); the feeling that goodness is no longer possible in the world (but might have once been before the violence); and a feeling that the world is revealed as never having been good at all (good is an aberration, if real at all). These are all distinct, although related, responses to extreme political violence. They are related because they all assume that the way one perceived the world and one's self, particularly in terms of morality, is no longer accessible. Something has gone wrong and there is a break in moral ability and the worthwhile nature of effort.

These relations to the *Good* remain distinct, however, and in important ways. If one feels that they are no longer able to be good or worthy of goodness, yet that others might be, shame will be a strong feeling. One may take on the burden of the demise of goodness themselves. If one believes that goodness has been eliminated from the world, or one's corner

---

[39] Again, this does not represent all experiences of war. We are, instead, focusing on a narrower set that comes from specific accounts.

of it, great sorrow can result, and there may also be remorse that one's community or people had a hand in it. We can think here of the quotation earlier in which the individual regrets the fact that his people have become thieves. One can feel that they have betrayed something important in the world. If one feels that violence has revealed a cruel world that has always been so, one can feel that they have been betrayed, leading to bitterness and resentment. Guilt, shame, remorse, regret, anger, and a sense of betrayal—all of these can color one's vision of the *Good* and raise doubts about its ultimate efficacy or even if goodness is possible.

When this happens, Murdoch's representation of moral subjectivity provides us with an image to visualize the change. The snap or slack means one is no longer living into the different modalities whose negotiation makes up a moral life and, to a substantial degree, our personality and identity. Another way to put this is, that the modalities of moral subjectivity that comprise the ways we search for and orient ourselves to the *Good* weaken or no longer seem intelligible to us. The gravity of void as constant potential negation remains, however, and we sink toward an experience of meaningless, despair, anger, and so on. The extreme suffering and loss of self undermine one's felt ability to move toward the ideals and images and obligations embodied in the other modalities. It is not that we do not feel able to be good; we do not, however, necessarily feel able to try to be good. Doubting that goodness is possible, or that it ever existed, we search our local worlds with tools we feel to be useless, trying to find something we doubt exists or that we feel has passed forever beyond our sight. It becomes possible, then, to drop our tools and give up the search. One's moral subjectivity is dominated by void, as the other modalities, as well as being a moral subject capable of goodness, slacken into inertia and inactivity.

If we refer back to Maček's account, we can illustrate these dynamics by interpreting the comments of her informants through this understanding of subjectivity. First, the cherished horizons embodying goodness as they understood it became less and less plausible for many. It was harder to move toward those horizons, as the ends they represented—say, being a good spouse or parent, some vocational dream, being a good neighbor, achieving great things—became implausible, or at least strained, in the newer context of war and social upheaval. Survival, for many, grew quickly in importance. This occurred not merely because of a desire to continue living. In a war where one's people and culture, or both, are being eliminated from a certain geography, survival becomes a critical form of resistance to genocidal coercion. This is particularly true if there is a great imbalance in terms of arms and power between adversarial groups. To survive, in other words, is to thwart more genocidal agendas. By surviving, one can in many instances carry on memories and culture, as well as one's existence, which allows the possibility, in the future, of renewal.[40]

Survival, though, can come at a price. Ends, goods, and needs, all embodied in the imagery of these horizons, compete, and, as we shall discuss shortly, it is difficult to hold these in tension. Survival, for example, often means putting aside horizons that were a priority in one's previous life, horizons in which one's identity as a moral subject was wrapped. We can think here of the image of oneself as trustworthy, dependable, a good neighbor. This could also include an understanding of oneself in relation to God, in which there is need to love others, to see them in the most charitable light, and to care for others as children of God. Putting others first, however, is not always conducive to survival. One may have to steal. One may have to hate and resent those who are firing on their home. Anger, ideas of vengeance, can become motives that fuel the fires of the survivor. Killing, manipulating, and violating one's sense of self as a morally good person, or one able to be good or even worthy of goodness, becomes imperative. As Victor Frankl, a psychoanalyst and Holocaust survivor, put it, referring to that previous genocide,

> … there was a sort of self-selecting process going on the whole time among all of the prisoners. On the average, only those prisoners could keep alive who, after years of trekking from camp to camp, had lost all scruples in their fight for existence; they were prepared to use every means, honest and otherwise, even brutal force, theft and betrayal of their friends, in order to save themselves—the best of us did not return."[41]

VISION, VOID, AND VULNERABILITY

One's moral orientation and ability are harmed through experiences of violence, as the experience and the knowledge of one's self or the world that it affords slackens or snaps our orientations and experience of value in the world. Yet, we can say something more about how this works, drawing further on Murdoch's understanding of virtuous vision and returning to the previous discussion on vision and vulnerability in Murdoch's thought.

---

[40] For an example of situations in which survival is not enough to continue cultural norms and practices, see Lear, *Radical Hope*.

[41] Frankl, *Man's Search for Meaning*, 19.

I have focused on Murdoch's understanding of moral subjectivity as it is a relatively understudied area of her work that has the potential to describe moral experience. There is not, however, a clear connection between her discussion of the moral modalities and void and her understanding of vision as the central metaphor and activity of moral development and transformation. Murdoch does, however, state that what she has written about in *Metaphysics as a Guide to Morals* mostly concerns Eros, and in that statement, we have a way to integrate more fully this representation with the rest of Murdoch's work.[42]

Again, Eros "is the continuous operation of spiritual energy, desire, intellect, love, as it moves among and responds to particular objects of attention, the force of magnetism and attraction which joins us to the world, making it a better or worse world …"[43] This description encapsulates much of Murdoch's discussion of the moral life, including attention as a central moral practice and metaphor, as well as the energy—the impetus—necessary for the moral effort needed to develop as a good person. Murdoch emphasizes the need for duty and a place for the will to describe aspects of moral experience that Eros and virtue do not capture.[44] Likewise, she includes axiom to make sure that the political aspects of life, particularly as she had lived during a great period of conflict between political ideologies, are represented in our experience, as they too can become elided in more deontological and erotological descriptions of the human. This representation of moral subjectivity at the end of *Metaphysics as a Guide to Morals*, then, is a sort of summary, as Antonaccio claims, in that Murdoch reveals where the different conversations and concepts she had been developing fit within the actual experience, which is the central background against which she philosophizes.

Murdoch's discussion of attention, as well as imagination and metaphor, then, are relevant for a specific aspect of the moral life, even if the erotic is, as she says, most of what we think of when we think of morality.[45] Within this discussion, Murdoch had, as we mentioned, an understanding of moral development that involved centrally an understanding of moral perception, in which such development consists in cultivating the ability

[42] Murdoch, *Metaphysics as a Guide to Morals*, 497.

[43] Murdoch, *Metaphysics as a Guide to Morals*, 496.

[44] This inclusion of duty, found in *Metaphysics as a Guide to Morals*, is largely missing in *The Sovereignty of Good*, most likely as Murdoch's emphases in the latter were to push back against philosophies she saw as dominant at that time that emphasized such terms.

[45] Murdoch, *Metaphysics as a Guide to Morals*, 497.

to see the world more and more accurately. There is in this a notion that there is something more accurate, which brings up the possibility that Murdoch is proposing a reality, perhaps a metaphysical plane called the *Good*, that we perceive and represent through our perception. Such a reading, however, is not necessary. As some interpreters have argued, her conception of metaphysics is much more modest.[46] Real is a moving target that describes the vision that is less and less self-referential, self-centered, and that is more accountable to the lived reality of others, their experience, and the material and spiritual conditions of their lives. Our vision is more accurate, in other words, the more we approach a perspective on the world where we are decentered, as this reflects some fundamental truths in the world. We accomplish this, in part, by focusing charitably—lovingly, Murdoch writes—on others. This brings us away from a world where we are at the center and converts us to a more and more accurate world of countless subjects with no real center. This also means that we are always developing a worldview that Murdoch compares to a constantly receding horizon of goodness. It may be more accurate to say a "constantly shifting horizon," as our ability to see is changing with experience and effort, as well as one that is never quite accessible. We are, in other words, never fully able to perceive a world in which we are totally decentered.[47]

What this means for the present inquiry comes into view when we remember that any such sight is neither permanent nor perfect. During war, the horizon of one's moral vision may transform quickly and enduringly to see one's own loved ones or group much more favorably, and others, such as strangers, uncharitably. The circle of care may narrow and close, and although one may be very generous and good within that circle, one may be vicious to those without. This is a fundamental transformation

[46] See especially, Broackes, "Introduction."

[47] Robjant and Conradi see Murdoch as arguing that we can sidestep, not end, our egocentrism (Conradi, "Divine though Unfinished," 30; Robjant, "Who Killed Arnold Baffin," A181). This seems right, but at the same time Murdoch articulates an ideal in which one would so develop their moral vision that, given a situation, one would not need to deliberate but, instead, the necessity of what one perceives would make action clear. This would imply that there is an ideal where there is no tension, where there is no competition between goods and images of the human, and where egocentrism has been tamed or made inert. I take this, however, not as a way of being Murdoch thinks will or can actually happen, but an ideal to clarify the nature of the *Good*, as well as an ever-receding ideal that we can strive toward (Murdoch, *Metaphysics as a Guide to Morals*, 331). At the same time, this does raise questions about Murdoch's ultimate hopes for the moral life and what is actually possible, and opens opportunities for different interpretations of Murdoch's thought than I take up here.

both in moral subjectivity and of one's local moral worlds, as the constituents of the local change, narrow, become less generous. The transformations go hand in hand, limiting the value one accords others as one limits one's network and community. We can lose our previous perspective, then, and begin to feel that too charitable a vision of others comes with risk. Loving attention, in this understanding, must be practiced sparingly. One may come to feel a moral phenomenology of mistrust is needed. And one's worldview and ethos will provide an epistemological, maybe even ontological, ground and rationale for valuing one's own group and disparaging another.

During war, this may actually be a correct understanding of what is happening and what the risks are of trying to live into a more ideal, more expansive understanding of goodness. What it means, however, is that one's worldview will change through violence, and, according to Murdoch, this means that one's subjectivity will also change. As one's subjectivity changes, so does one's identity and how one perceives oneself and others. Love broadly hurled can become dangerous and may be replaced, understandably, by attention that is more the gaze of the predator, or at least the wary, than that of the charitable. And thus, the sometime linkage of the saint and the fool.

This is not Murdoch's own understanding, yet I argue that political violence, with which she does not directly contend, has this interpretive effect and requires further extension of her thought. The fact that one can come to feel less generous and may come to feel vicious toward others is a reflection of the way in which one's surroundings have constrained choices and made certain options or subjectivities very costly or near impossible. Virtue ethicist Rosalind Hursthouse writes that eudaimonia may not be possible until better days come.[48] Yet, the situation may be even more extreme than this. As philosopher Lisa Tessman argues, environments of oppression can create selves with different virtues aimed at goals that differ from those people and groups who are not living under domination. Such subjectivities and the burdened virtues they include will be enduring, if not permanent.

In the case of more explicit violence, while we wait for better days in which eudaimonia may be accessible, we change in order to get through the present violence. Subjectivities are not a light switch, however. As anthropologist Veena Das argues, our former worlds haunt us in the form

---

[48] Hursthouse, *On Virtue Ethics*, 177.

of impossible values and ideals.[49] One can go through violence, transforming their moral subjectivity to work toward survival of themselves and a more circumscribed group. This moral subjectivity will continue into a new, post-war, post-conflict world. The impression or echo of that subjectivity developed to survive violence will bear the imprint, habits, and virtues of that time into the new world. One will continue to change, but the specter of past experience, and its influence on one's soul will remain to some degree.

One's ends, that is, one's telos and moral horizons, change. This change raises the question of whether *the Good* itself changes. From the point of view of the individual's experience, it can transform, become insurmountably distant, or disappear. Survival, as we have already discussed, can take the place of other ends, and so will require different virtues, different ways of being a moral subject. In addition, one's moral vision will also change. For example, the experience of violence and the new landscape that violence creates will change the way one sees. If we take seriously, as Murdoch does, that cognition is based in imagination and metaphor, and if we see metaphor as relating two separate objects or ideas to create something new, we must take seriously that during times of violence, the objects used to make such metaphors will change, and the contents of the metaphor may create thoughts and worldviews based in images and experiences of sorrow, betrayal, disappointment, and loss.[50] This can change the way one sees the world, just as it changes the experience of tension in the moral life, leading one to despair of living into certain cherished images and norms, just as one is no longer able to see the world as they once did.

This is a transformation of the aspects of subjectivity and of identity, a transformation of one's self, and yet parts of the old self live on. As Veena Das has insightfully written, one may find themselves in a new world and new circumstances yet still remember past norms.[51] Individuals, institutions, and communities may still hold one to those norms, even if the elements that made their realization and pursuit possible have vanished in violence. There is a spectrality in the moral life in which who one was

---

[49] Das, *Life and Words.*

[50] "The development of consciousness in human beings is inseparably connected with the use of metaphor. Metaphors are not merely peripheral decorations or even useful models, they are fundamental forms of our awareness of our condition: metaphors of space, metaphors of movement, metaphors of vision" (Murdoch, *The Sovereignty of Good,* 75).

[51] Das, *Life and Words.* See especially Das's discussion of "Asha" in Chapter Four, pp. 59–78.

haunts who one has come to be. This is one of the ironies of pursuing the *Good*, one of the side effects of being a subject who reaches toward the Good: being good means one is more susceptible—more vulnerable—to situations where one is unable to live into past morality. One holds oneself accountable, even if one no longer has power to follow past moralities. The past haunts the present, one evaluates their present self-based on the past, and this can lead one to feel even more that they are far from being good. The road back to orienting oneself toward goodness is felt to be too far, too inaccessible, or too deeply buried in the dust. The structural vulnerabilities of the moral life I discussed earlier are the source for a continued, possibly toxic, responsibility.[52]

What I have presented is a sketch of an understanding of moral subjectivity interpreted through Murdoch's thought and understanding of virtue. The moral-philosophical anthropology she creates allows us to show what it is about one's subjectivity that changes during the violence, as well as to provide an account that, if brief, allows us to discuss how those changes come about. Political violence can undermine the ability of virtue to want to move toward the *Good*. One becomes literally disoriented as one's local moral world collapses and as one's moral subjectivity is called into question, as the elements that condition one's teleological drive to orient oneself toward the *Good* are challenged or weakened. Such lived teleology allows us to create relationships and community by pushing us to strive toward living into images of the human, and away from others so that one can have the virtues, especially social virtues, that are the keys to others' acceptance and even embrace of us. Trustworthiness, courage, generosity, as well as other virtues, are valued in friends, lovers, co-workers, compatriots, and we reach toward those virtues and the images of the human they are a part of not just because this allows for the survival of a social animal but also because it allows for community itself and for a life that is meaningful, one where one might flourish. All of this happens through the metaphoric nature of thought, through the structure of vision and perception, and the ways we change as the world changes around us.

Such abilities and the foundation of enabling communities may, then, be more vulnerable than they seem, as the tension that characterizes moral subjectivity can collapse from extreme violence and suffering. This means that Simone Weil's claim that "all sins are attempts to fill

---

[52] See, again, Sherman, *Afterwar*, 72–104.

voids" is the inverse of what I have described here. For Weil, who had a strong influence on Murdoch, spaces of spiritual void in the end affirm through their absence God's inevitable presence. Voids in our examples are not filled by sins. The necessity to act against one's moral values and against the visions of the human one hopes to embody creates, if not the possibility of void, then a moral subjectivity dominated by void. One can experience this as sin, but it is in reaction to the radical transformation of one's local moral worlds, one's relationships and values, and the routines and structures that make up our existence and its context. Such elements we take for granted, and when they are revealed through violence they are simultaneously damaged or eliminated through the same violence. One's world can collapse.

Even though I admire Murdoch's emphasis that goodness remains, the devastation that human beings can inflict upon each other will insist on having conclusions of their own. The extreme severity of suffering that sometimes defies both meaning and hope is not something that can be responsibly argued away. For some, and in some situations, the *Good* will never shine brightly enough to call one from the shadows. There is a point at which theodicy must fall silent and hang its head. Grace in the form of unexpected capacities that emerge seemingly without having been earned may break through formerly unsurpassable obstacles, and we err when we too readily bound the resilience of human selves and communities. Yet we also err if we do not make room for despair's insistence, and to recognize that sometimes despair has the final word. Not just grace and gravity, but grace and utter void are both potentials in human life. When we do not take account of these aspects, our images of the self and society become inhuman, and so inhumane in that we limit our ability to recognize the reality of our existence.

Political violence, then, is moral violence, and we miss something when we do not attend to the ways in which such conflict transforms our sense of self and the way we see the world. Murdoch is insightful here, as she calls attention to the way that vision is inherently evaluative. She does not mean this just metaphorically. Moral judgment exists in the organs of the eyes as much as in the lobes of the brain. If we take a moment to pay attention to the way we engage the world through vision, we are constantly evaluating and categorizing what we see and dealing with the feelings and thoughts that arise from our perception. This is even more true in situations of existential risks, such as political violence, as what one sees is constantly changing, and where one even begins to see loved ones in a new light.

In such situations, there may be no good choices even as one must act within situations where the stakes are high, where loved ones' lives are at stake, where one feels that the outcomes of one's actions will affect one's soul. If the new world that arises, one of violence and devastation, constantly admits of no good choices, the effort to be good can feel futile. One's moral abilities can seem to have disappeared. Murdoch's hoped-for ideal—where we become such perfect moral perceivers that choice is not necessary—meets its inverse in extreme violence, where there are too many choices. In periods of extreme violence, the possibility of one right choice recedes and is negated.[53] There is collapse and flatness, and we can no longer turn toward the sun or are no longer sure whether the shadows we see are cast by a sunrise or a burning world behind us.

---

[53] This is not an argument for MacIntyre's critique of modernity as fracturing and morally unintelligible, grounded in an understanding of *tradition*. Instead, it is an argument about how political violence can make for extreme situations where no good options exist. Such conditions can exist in any period and within any tradition. In addition, Weil is important in this discussion, as Murdoch seems to cite Weil as the source of her idea that perfected moral vision will eliminate choice and would result in the necessity of specific actions (Murdoch, "Knowing the Void," 159).

CHAPTER 6

Moral Subjectivity, Moral Injury

For some of those who survived the war and the siege of Sarajevo, the experience left a lasting mark on the soul, to put it poetically. One may survive only to feel they have been altered, perhaps wounded, in a way that has transformed negatively their ability to relate to others and even to have hope in one's own goodness, than that of others, or that of the world. If one's subjectivity becomes dominated by void for long enough, or deeply enough, it can become a morally injured subjectivity. Indeed, the length of time is not necessary for such a transformation. Violence has a way of compressing time and grabbing control of the shape of one's future. Domination by void over a long period can change one's attitude, one's outlook, habits, and desires, yet intense, violent events or environments can empower such domination to leave a durable mark. Even when one continues to live long past the period or ecology of violence, and even when one returns to peacetime or to one's home, a morally injured subjectivity can make it hard to inhabit new worlds, new relationships, and new subjectivities, or in the case of soldiers, and those returning home from war, old worlds, and old relationships. Such experiences can so transform one's sense of their own abilities and identity, as if one has been misshapen and can no longer fit through human-shaped doors. They may feel inhuman, or less than human, or not able to re-enter the world of civilized humanity.

We return here to the language of the monstrous. One can feel ill-equipped to live in a world or a society defined through peacetime norms when violence has so shaped the possibility of future subjectivities. Part of

© The Author(s) 2019
J. Wiinikka-Lydon, *Moral Injury and the Promise of Virtue*,
https://doi.org/10.1007/978-3-030-32934-1_6

such an injury comes from the violence and from the experience of war or genocide. Another part, however, comes from comparison during or after such experience when an individual compares herself to people perceived as more "normal," untainted, who live seemingly peaceful and easily in societies that war or genocide have not embroiled. This is, after all, the language—of "normality"—that Maček's informants use to articulate their anxiety that they are moving away, perhaps irreparably, from the norms of peacetime European life. Moral injury, then, is at least twofold in terms of experience in that one is injured not just by the violence itself and the way it shapes one's surroundings, which in turn shapes one's consciousness. It also comes from the way one sees, the way one compares and evaluates the lives of others, and how one feels they hold up against such a comparison. It is as if, to put it in Murdochian terms, one is unable to attend lovingly to one's own experiences to put them in proper perspective and unable to see the contours of another's everyday life.

What can make this worse is that others can also see the morally injured as monstrous. For example, U.S. veterans returning home from wars and who cannot seem to reacclimate to peacetime culture can soon be seen as dangerous. They can also seem to be distortions of their former selves. Once good, partners or parents, they can seem to be different people, a different spirit inhabiting the same body. What was a source of love and life, such as one's children, can be seen as dangerous, as a child's energy can trigger post-traumatic symptoms such as hyper-vigilance.[1] And for entire populations, such as in Bosnia, other countries can see them as now outside the fold of civilized nations, of Europe, as attested by the admission of former Yugoslav states such as Slovenia and Croatia to the European Union, but not Bosnia. There are economic and political factors involved in such decisions, but there are deeply moral, even phenomenological reasons, as well. Others see Bosnians not only as needy and a potential drain on their country's resources but also as sources of violence themselves. Some accounts see this as a form of contagion, yet this is less an issue of contagious mental energy than it is of responses to intersubjectivities that are morally harmed and seem to be unfit or unable to re-enter intersubjectivities of (seeming) peace.[2] This, then, is the third source of moral injury,

---

[1] Benimoff and Conant, *Faith Under Fire*, 84, 203.

[2] In speaking of contagion, I am thinking of neo-Freudian theories, such as those of René Girard.

where others seem to confirm the monstrous nature of one's moral reformation under violence.

I have framed moral injury in terms of subjectivity, as harm is done to
one's ability, or at least one's perception of their ability, to live into or
imagine certain moral intersubjectivities or the very possibility of such
intersubjectivities. It affects perception and imagination, of how one sees
the world and themself and how one can imagine different worlds, different selves, and different relationships. This, however, is not a typical framing of moral injury. The understanding of what exactly moral injury is,
what exactly is injured, and why such injuries should be considered
"moral" ranges widely. Some see such injuries as damage inflicted on an
individual, as a form of victimization, while others use moral injury to
articulate harm done to those who inflict harm, a word for the dehumanization of violent perpetration. In the context of war, however, categories
such as "victim" and "perpetrator" easily fall apart. Survivors can feel
complicit; perpetrators and soldiers can feel like victims. Indeed, whether
an objective observer assures one of their virtue, one can still *feel* both
victim and perpetrator. And some of Maček's informants illustrate this, as
their attempts to survive war, genocide, and siege can make them feel part
of the war, instead of a victim of it.

The understanding of moral injury presented here is distinct from these
other approaches or discourses in that it shares commonalities with each
yet fully belongs to none. It aims at an understanding of moral injury that
draws from insights of various approaches not only to better reflect the
experience of Dizdarević and others but also to demonstrate the importance of engagement between these approaches. There are roughly three
distinct, existing approaches with their own origins, what I refer to as
"clinical," "juridical-critical," and "structural" moral injury discourses.
This work ends with a look at these understandings of moral injury to better locate and articulate the type of harm I have been discussing in these
pages by engaging these discourses on moral injury. What I have argued
so far is that the harm that Dizdarević and Maček described is a harm to
one's moral subjectivity, the harm described by domination of what
Murdoch called "void." This is a lasting harm, however, even an injury,
but even more, such experience is injurious to one's life. Morality itself can
look unintelligible or even laughable as one's perception of the world may
have been so altered that the world, that existence, is not a place that can
nourish goodness. Indeed, one can even come to see the world as mock-
ing such attempts. It can become harder because of such experiences of

violence and war to relate to others, to try and inhabit peacetime subjectivities, or imagine different futures for oneself, one's loved ones, and one's community.

It makes sense, then, in the final chapter of this work to locate the harms described in previous chapters, harms that are moral in a profound sense, within discourses on moral injury. This should not only expand the vocabulary that can be used to articulate the harms that Dizdarević and Maček are witness to. It should also help expand current notions of moral injury to reflect a wider range of experiences in war.

CLINICAL DISCOURSES OF MORAL INJURY

Perhaps the best-known use of the term "moral injury" comes from the fields of psychiatry and psychology. Psychiatrist Jonathan Shay is usually credited with coining this use of moral injury in the early 1990s, beginning what could be called a clinical discourse of moral injury. Created in response to the experiences of U.S. veterans of the Vietnam War who were under his care and whose suffering was not wholly explained through trauma or PTSD discourse, Shay originally thought of moral injury as the result of a "leader's betrayal" of "what's right." As he has developed his definition, Shay has argued that "moral injury is present when (1) there has been a betrayal of what's right (in the soldier's eyes) (2) by someone who holds legitimate authority (3) in a high-stakes situation."[3] The term itself seems to be one that Shay developed from whole cloth, having developed the language of "injury" in response to the realities of U.S. military moral culture, based as it is on discourses of duty and virtue. Shay's goal was to replace the psychological terminology of "disorder," as is found in diagnoses such as "post-traumatic stress disorder," with "injury," to signal that mental health wounds do not reveal a mental or characterological deficiency but instead represent an injury as serious and noble as any sustained in combat.[4]

A decade and a half on from his original use of the term in 1991, other psychologists, what could be called the second wave of clinical moral injury discourse, have taken up Shay's understanding of moral injury, which is in the process of being rationalized into a standard diagnosis.[5]

---

[3] Shay, "Casualties," 183.
[4] Shay, "Casualties," 181.
[5] See specifically Litz et al., *Adaptive Disclosure.*

These newer approaches, many of which do not reference Shay, nevertheless agree with him that the experience of veterans and soldiers requires more nuanced than the current vocabulary of trauma affords alone.[6] Their focus also remains practical. The goal generally is to understand moral injury as a form of harm, to legitimize moral injury diagnoses by identifying cases of moral injury in previous conflicts where it went undiagnosed, and ultimately to create therapies to alleviate such suffering.

Although sharing several assumptions, the second wave of clinical discourse diverges from Shay in several ways. The subject of moral injury for clinical discourse is understood as a veteran soldier—in Shay's case, specifically a U.S. veteran in the context of war. While agreeing on the subject of moral injury, there is disagreement at times around what or who commissions moral injury. For later clinicians, such as Bret Litz and a number of psychologists who have worked with him, there are at least two possible sources of moral injury: the soldier herself and the environment of combat. In the first case, the injured is harmed because she has committed actions or participated in situations that violate her moral code. She is both recipient and source of that injury, both subject and commissioner. This is not to say that Litz and his co-authors neglect other factors; it is to say that the soldier is a central agent in her own moral injury. In addition to the soldier being such a reflexive morally injured subject, she can also become morally injured from seeing atrocity or handling the remains of the dead.[7] In such situations, one's moral injury does not arise from any direct violence, nor can it be said that a soldier commissions their own moral injury. Instead, the assumption seems to be that it is the war environment itself, or at least one's perception of it, that gives rise to subjectivity of moral injury. No direct violence is involved, yet moral injury remains an aspect of the general environment created by the violence of war. This differs from Shay in that for him the soldier remains the subject of moral injury, but, as I illustrated previously, the commission of moral injury rests in the hands of superiors and even in the military hierarchy itself. While other clinicians might see moral injury as personal or even environmental, Shay sees it as an institutional issue.

Moral injury has been since its inception a term with an inherent critique. One such critique is shared throughout the clinical discourse. For example, the above authors share a critique of the previous way that

---

[6] Nash and Baker, "Competing and Complementary Models of Combat Stress Injury," 75.

[7] Litz et al., "Moral Injury and Moral Repair in War Veterans," 696.

trauma was understood and deployed in psychiatric and therapeutic practice.[8] Clinical discourse arose as a response to the dominance of trauma and PTSD as a way to understand and treat the mental and emotional suffering of veterans—trauma and PTSD themselves having arisen as a critique of previous approaches to wartime psychological distress. In the continuing search for concepts, frames, and approaches that better reflect the reality of mental, emotional, and spiritual harm resulting from war and political violence, clinical discourse stands on the assumption that PTSD is insufficient in this regard.

Shay's institutional emphasis, however, raises a second, inherently political critique that is lacking in other clinical writers.[9] Since moral injury comes from betrayal within the chain of command, commission rests in the hands of a military institution. This has a direct effect on what constitutes repair. For the second wave, repair involves the therapeutic treatment of an individual, or sometimes a small group of individuals, under the care of a licensed clinician. Shay's emphasis on the institutional context of moral injury means that moral or "soul" repair needs to come not just from individual or group therapy but also from the military in terms of reform.[10] Repair is understood, then, not just as therapy but also in terms of education, advocacy, and institutional reform. Later, clinicians lose this critique, as their understanding of moral injury focuses more narrowly on the individual soldier rather than on the political and institutional context in which they were injured. This may seem surprising, as the acknowledgment, noted previously, that the environment of war creates moral injury may in itself seem a sufficient basis to critique the practices and institutions of war.

There are exceptions, however, such as Tyler Boudreau, who has recently used moral injury to criticize the injustice of wars such as those in Afghanistan or Iraq. Boudreau, however, is a veteran and former marine officer and not a therapist. He seems to draw his usage from cli-

---

[8] For others, see Drescher, et al., "An Exploration of the Viability and Usefulness of the Construct of Moral Injury in War Veterans," 8; Steenkamp et al., "How Best to Treat Deployment-Related Guilt and Shame," 471; Nash et al., "Psychometric Evaluation of the Moral Injury Events Scale," 646; Litz et al., "Moral Injury and Moral Repair in War Veterans," 696. For an argument that PTSD and related treatments are sufficient, see Smith et al., "Perceived Perpetration During Traumatic Event."

[9] I discuss this inherent critique more deeply in Wiinikka-Lydon, "Moral Injury as Inherent Political Critique."

[10] Brock and Lettini, *Soul Repair: Recovering from Moral Injury After War.*

nicians associated with the Department of Veterans Affairs, and his emphasis on seeing foreign policy as central to moral injury is more in keeping with Shay's original dual critique.[11] In this way, he provides an alternative to current understandings of moral injury in psychology that do not insist on addressing the structural or contextual sources of such distress. Indeed, in so far as such clinicians can be said to have a concern with institutional context, it is limited to that of the institutional role of the combat soldier.

Another key way that Shay's original formulation differs from the later discourse is his understanding of betrayal, although the daylight between these positions is narrowing. Second-wave writings focus on the guilt and shame of the individual soldier, and so, on what can be called "self-betrayal."[12] The definition originally developed by Litz and his co-authors in their seminal article, "Moral Injury and Moral Repair in War Veterans," which has been influential in studies of combat and mental health, stresses such an understanding of self-betrayal when it defines "potentially morally injurious events" as "perpetrating, failing to prevent, or bearing witness to acts that transgress deeply held moral beliefs and expectations."[13] It is the soldier, then, that betrays her own moral commitments, and so is the agent or commissioner of the moral injury.[14]

Shay, in contrast, originally viewed moral injury as the result of betrayal by a superior within high-stakes situations, which would put him at odds with the second wave. He is joined here by philosopher and psychoanalyst Nancy Sherman, who sees the moral injury, at least in part, as resulting

---

[11] Boudreau, "The Morally Injured," 2011, 748.

[12] See Litz et al., "Moral Injury and Moral Repair in War Veterans," 698; Stein et al., "A Scheme for Categorizing Behavior Modification;" Smith, Duax, and Rauch, "Perceived Perpetration During Traumatic Events;" Keenan, Lumley, and Schneider, "A Group Therapy Approach to Treating Combat Posttraumatic Stress Disorder;" Bryan et al., "Moral Injury, Suicidal Ideation, and Suicide Attempts in a Military Sample;" Sherman, *Afterwar*.

[13] Litz et al., "Moral Injury and Moral Repair in War Veterans," 695. For studies drawing on Litz et al. and accepting their definition, see Worthen and Ahern, "The Causes, Course, and Consequences of Anger Problems in Veterans Returning to Civilian Life;" Paul et al., "Prolonged Exposure for Guilt and Shame in a Veteran of Operation Iraqi Freedom;" Vargas et al., "Moral Injury Themes in Combat Veterans' Narrative Responses from the National Vietnam Veterans' Readjustment Study;" Smith, Duax, and Rauch, "Perceived Perpetration During Traumatic Events."

[14] See also Drescher et al., "An Exploration of the Viability and Usefulness of the Construct of Moral Injury in War Veterans," 9; Steenkamp et al., "How Best to Treat Deployment-Related Guilt and Shame: Commentary on Smith, Duax, and Rauch," 472.

from orders forcing one to betray others, such as fellow soldiers under one's command or civilians.[15] For Sherman, even though the order comes from another, the soldier's action in carrying out that order internalizes responsibility and guilt. Shay has, however, recently acknowledged the possibility of self-betrayal, making his understanding broader and more nuanced. At the same time, Litz and his co-authors seem to have acknowledged betrayal by others as potentially significant.[16] This is creating a broader understanding, still being developed, of the experiences and relationships in which moral injury manifests.

JURIDICAL-CRITICAL DISCOURSES OF MORAL INJURY

A separate discourse defines moral injury not in terms of the commissioner's experience but in terms of the subject. This discourse draws from various sources. Typified by philosophers Jeffrie Murphy and Jean Hampton, one approach focuses on Bishop Joseph Butler's moral philosophy. Murphy and Hampton, in particular, developed an understanding of moral injury, based on Butler, as a way to help clarify the stakes involved for the philosophy of law when looking at the tension between forgiveness, mercy, or retribution.[17] Another approach draws from critical theory, particularly Axel Honneth's understanding of moral injury as the result of misrecognition. More recently, J.M. Bernstein has drawn from both critical theory and Hampton's more juridical use of moral injury to develop an understanding of moral injury as a form of devastation that results from extremely dehumanizing violence, such as torture and rape. What these approaches have in common, other than their focus on philosophy as opposed to psychology is an understanding of moral injury as dehumanization. This discourse, which I call "juridical-critical," separates sharply the commission and subjectivity of moral injury, where one person

---

[15] Sherman, *Afterwar*, 77–9; Shay, "Casualties," 183. The political critique of Sherman's writing, however, largely focuses on raising awareness about the suffering of veterans and the public's responsibility in the moral repair of veterans (*Afterwar*, 40).

[16] See the differentiation between "Moral Injury by Self" and "Moral Injury by Others" (Nash et al., "Psychometric Evaluation of the Moral Injury Events Scale." The authors do not, however, address whether this categorization might change their theory based as it is based on a victimization/perpetration binary (see also Stein et al., "A Scheme for Categorizing Behavior Modification").

[17] Murphy and Hampton, *Forgiveness and Mercy*, 15–16.

commits violence on another.[18] This makes the juridical-critical discourse the mirror opposite of the second-wave clinicians, at least in this regard. There is also no special context or institution in the juridical discourse, such as combat zones or the military. Instead, moral injury can result from various contexts, from daily social interactions to extreme cases, such as rape, torture, and hate crimes.

Concerning issues of moral injury and forgiveness, a key text is Murphy and Hampton's *Forgiveness and Mercy*, whose usage of moral injury predates Shay's use by about 3 years. Written as a series of essays in which the authors debate one another, back and forth, Hampton and Murphy's main concern is with the appropriateness of retribution, hate, and forgiveness in legal philosophy and jurisprudence. The relevance of their exchange for the present work, however, lies in their understanding of the moral injury. Invoking Butler, Murphy understands moral injury broadly as a wrong that harms another's self-respect. He argues that such wrongs count as moral injuries either when someone's rights are violated or when another is caught "free riding," thus taking advantage of the fruits of another's work and sacrifices. Freeriding is a moral injury, for example, because the very act communicates a valuation, or really a devaluation, of the injured. One's rights or one's work are not worth being respected, and so the individual herself is disrespected and devalued. Such wrongs can merely be experienced as an insult, but in more significant cases they can actually degrade one's felt value. Indeed, Murphy argues that such injustices are usually betrayal, creating a similarity with Shay's definition, although betrayal for Murphy can occur beyond the military context.[19] By understanding moral injury in this way, Murphy gestures toward a much broader range of actions and relationships than Shay or the clinicians acknowledge. Moral injury is not just about violence but instead occurs in everyday actions in everyday society.[20]

---

[18] As I will demonstrate when discussing Bernstein's work, Bernstein draws from recognitive theories specifically, as well as others. The label "juridical-critical," then, reflects some, but not all, of the key sources of these theories.

[19] Murphy and Hampton, *Forgiveness and Mercy*, 15–17.

[20] Murphy does not use any examples, however. What separates a wrong that is a morally injurious from one that is not, then, is not clear. Indeed, there is an understanding here of time and effort as media of value, where one's rights and energy are not worth one's effort, revealing and so communicating implicitly the agent's belief that their activities of energies are more valuable. See also Hampton, "A New Theory of Retribution," 387. Murphy also seems to imply that moral injury is a key constituent of the social and of subjectivity, since

Hampton agrees with Murphy that moral injury can arise from everyday interactions and disputes, such as those with family, friends, and roommates.[21] This does not mean that for either Murphy or Hampton all such injuries are therefore equal. There are more or less significant degrees of injury where one can be demeaned or, more seriously, degraded. The more significant moral injury involves what Hampton calls "diminishment."[22] This can be revelatory, when the injured believes the insult or demeaning behavior reveals some truth about them, namely, that they are of lower value than they had once thought. In such a case, the injured re-evaluates the way they previously viewed themselves, which now seems to have been incorrect or even delusional. Hampton aligns here with Murphy's concern with self-esteem, as the injury actively lowers self-esteem, thereby harming the injured in a psychological way. As bad as this is, there is also diminishment that is objectively degrading, in which one's abilities—not just one's self-respect—are actually lowered through the commission of the wrongful act. This can be relatively minor, such as a person losing the vocational ability—for example, when an athlete is physically hurt by another's depriving them of being able to play a sport. The athlete perceiving this loss as a fall from an ideal no longer attainable can be considered as a case of moral injury.[23]

Although such an injury and its meaning are horrible for the athlete and those close to her, Hampton stresses that there are wrongs that are more consequential for society more broadly. Writing after *Forgiveness and Mercy*, Hampton eventually defines moral injury more precisely in terms of "*dignity*."[24] For Hampton, dignity is socially defined and constructed.

---

such evaluations are made throughout the day and structure social interactions, connecting it with Gilligan's understanding of gender and moral injury, which will be reviewed shortly.

[21] Murphy and Hampton, *Forgiveness and Mercy*, 40.

[22] Murphy and Hampton, *Forgiveness and Mercy*, 49, 54. Hampton also talks about less serious, although still morally injurious, "demeaning" wrongs, where one does not feel they are permanently devalued by the wrong.

[23] Murphy and Hampton, *Forgiveness and Mercy*, 51. It should be noted that the value of being a mother or an athlete is culturally conditioned. The cultural definition of worth is key in Hampton's theories, as I will show in the following discussion on dignity in Hampton's work. Also key and implicit in the example of the athlete is that moral injury is intersubjective. It is a wrong done between subjects. It does not include more "natural" accidents. If, for example, the athlete fell, this would not be demeaning, because such wrongs do not emanate from the natural world, and so do not carry with them a communication of one's worth (Hampton, "A New Theory of Retribution," 394).

[24] Hampton, "Correcting Harms versus Righting Wrongs," 1684.

As norms surrounding dignity regulate how one is supposed to treat others, these norms are central to social cohesion. Consequently, harms that undermine those norms strike not just against an individual, but more significantly, against a society. Such an "*affront to the victim's value or dignity*" (Hampton's emphasis) is not just harm against an individual but an attack against the community.[25]

What gives a moral injury such social significance is, for both Hampton and Murphy, the communicative aspect of such harm. One may not be brought low by an insult or demeaning behavior, yet such actions do broadcast that another person regards you as having less value than society insists you have, or as occupying a lower rank than society thinks you deserve. This is morally injurious, then, because someone is putting forth an alternative theory of value in competition with the reigning theory of society.[26] This argument that moral injury has a central, communicative aspect raises and broadens the stakes of one's indignities. The more severe moral injuries are no longer just a personal matter but rather a threat to the standards that allow for peaceable social interactions.

The communicative dimension of moral injury means, however, that it is comparatively less important to define moral injury as something experienced as harmful or painful. Hampton acknowledges that such injuries are usually subjectively felt, yet the experience of the injury is less important to her argument than the injury's broader social effects.[27] What defines such a moral injury, then, is not subjective—that one feels injured—but rather objective, that is, how a third party would interpret the significance of the wrong. Its importance is social. To make this clearer, Hampton posits a hypothetical bystander, one who as a member of the morally injured subject's culture has internalized the same norms of human worth. If any comparable bystander views the wrong as violating another's dignity, then it is deemed a moral injury, communicating a theory of human value that challenges that of the society. The urgency, then, is not so much the individual's suffering, although that is not unimportant to Hampton. Instead, the urgency is in the social interpretation of wrongful actions and how they affect social norms and cohesion. If people are worthless, more people may conclude they can treat others unjustly or even cruelly,

---

[25] Hampton, "Correcting Harms versus Righting Wrongs," 1666.
[26] Murphy and Hampton, *Forgiveness and Mercy*, 52.
[27] Murphy and Hampton, *Forgiveness and Mercy*, 51–2.

weakening social trust and making society vulnerable through the increase of potential conflict and the vulnerability of its members.

Hampton argues that, given the stakes, society's answer to such a message should be forceful and unambiguous. The repair for such injury must communicate a triumph of the original social theory of human worth over the wrongdoer's challenge. Only retribution, according to Hampton, can convey through its violence the importance of human dignity and the seriousness of such infractions. Dignity is so valued that society is willing to, perhaps ironically, diminish that of the wrongdoer. The commissioner of the moral injury is thus defeated, even devalued, and along with her, her idea that another individual in society was worth less than she.[28] Retribution, then, is a form of moral repair for actions that degrade society through the degradation of individual human liberty.[29]

More recently, J.M. Bernstein has also conceived of moral injury in terms of dignity, social trust, and vulnerability.[30] Bernstein takes his cue from Hampton yet adds to it an understanding of moral injury drawn from critical and recognitive theorists, such as Theodore Adorno and Alex Honneth.[31] Like Hampton, and even Murphy, Bernstein views moral injury in broad terms that include everyday humiliation and betrayal, as well as physical violations that impinge on one's value as a human. Even so, the focus of his work is on extreme forms of moral injury caused by bodily, and even what could be called "spiritual," violence, if we follow Charles Taylor in understanding "spiritual" as the domain of human flourishing, meaning, and the conditions necessary to realize one's potential.[32] Indeed, while Hampton uses the language of "diminishment" to describe what is harmed in extreme cases of moral injury, Bernstein uses the

---

[28] Murphy and Hampton, *Forgiveness and Mercy,* 143.

[29] Hampton, "Correcting Harms versus Righting Wrongs," 1684. This does not mean that Hampton believes there is only one normative conception of the human in a society. She is concerned, however, more narrowly with retribution in the context of legal institutions, such as the courtroom, where possible normativity is more circumscribed by law.

[30] There are, of course, other philosophers that have also looked at humiliation and degradation in terms of agentive harm and breaking of world trust. See specifically Margalit, *The Decent Society;* Rorty, *Contingency, Irony, and Solidarity.;* Scarry, *The Body in Pain.*

[31] Bernstein, "Suffering Injustice," 304, 318; *Torture and Dignity,* 16–17. Honneth does not appear to use the term, but Bernstein argues that he is concerned about the ways that people are harmed through misrecognition, and that misrecognition is essentially a moral injury.

[32] Bernstein, *Torture and Dignity,* 4–5, 15; Taylor, *Sources of the Self,* 14.

stronger term "devastation," which indicates deep irreparable harm that, unlike Hampton, is necessarily subjective.

Specifically, Bernstein argues that devastating injury is possible because there are voluntary and involuntary aspects of the self. This means that, despite our sense that we are our own persons and in control of our bodies and responses, such control is limited. This involuntary side of human beings, such as pain reactions, can make us vulnerable if another person can take hold and violently manipulate those involuntary mechanisms. Bernstein is thinking primarily of torture and rape, which for him are paradigmatic examples of devastating pain. Both torture and rape are made possible by the fact that another person can, through violence, manipulate one's body against one's wishes, making one involuntarily scream, feel pain, move, and plead.[33] This loss of control can seem like one's own body has betrayed oneself by acting under the agenda, desire, and will of another person.

This claim has a broader social consequence. To get through the day, each individual must have a certain degree of trust that their neighbors will not devastate them. When such trust is broken, Bernstein is wary that it can ever be fully healed. After such devastation, it is much more difficult to ever have trust in a world where one could be so cruelly "undone." Neighbors, co-workers, even friends, are now seen as potential threats, and as a consequence, the world itself seems to be a ground of constant potential devastation, a place within which the morally injured subject will never feel completely at ease again.[34]

In describing the devastation, Bernstein follows Hampton in making dignity key to understanding the moral injury. For Bernstein, dignity is produced when one's community or society recognizes one's worth.[35]

---

[33] Philosopher Sarah Clark Miller anticipates Bernstein's understanding of rape as critical to understanding moral injury when she writes about rape and genocide in Darfur, also drawing on Hampton: "Rape serves as a quintessential example of a morally injurious action or behavior. Rape is an affront to a victim's dignity and an attempt to deny her equal moral worth. Through the act of rape, a rapist expresses a victim's inferior moral standing and his own corresponding superiority. Rape accomplishes the aims of diminishment particularly effectively, perhaps because of the intertwining of sexuality, humanity, and shame that it involves" (Miller, "Moral Injury and Relational Harm," 511). Bernstein takes his notion of voluntary and involuntary aspects of the self from philosopher Helmuth Plessner (Bernstein, *Torture and Dignity*, 198–208).

[34] Bernstein, *Torture and Dignity*, 13–6.

[35] Bernstein explores the recognitive aspects of his thought in an earlier article, "Suffering Injustice."

Here, of course, we can see his use of recognitive theory. To be human is to be recognized as such by other humans and to receive the benefits of such recognition. It is an intersubjective good. So, if dignity is understood as a status that one's community cognitively bestows on an individual, then moral injury results when one's dignity is deracinated through subject-devastating violence.[36] Devastation can do this because one's dignity, one's social standing, is a social institution used to safeguard one from the brute fact that others have the power to try and devastate one's soul. Dignity marks one as worthy of certain treatment and unworthy of demeaning treatment, while self-respect—what Bernstein refers to as internalized, felt dignity—makes one feel inherently worthy of such status. Devastation can remove dignity and one's self-respect by treating one violently in a manner beneath their worth and by making one feel they have been made to betray themselves. Of course, one's community can still say that the morally injured person still has dignity despite the violence. Bernstein argues, however, that the damage has been done. Drawing from Jean Améry's discussion of his treatment at the hands of torturers during World War II, Bernstein agrees with Améry that, since dignity is something bestowed, it is something that can be taken away. And it is just such devastating violence that can remove that feeling of security and trust.

Dignity, then, is not only socially important. It is also the recognition that although we need the social to thrive, it is this very condition that leaves us open to devastation. Because we are not in control of our entire embodied selves, others can take advantage of this to take control of us and harm us. Moral injury for Bernstein reveals dignity to be a uniquely critical recognitive, good quite critical to survival, and bodily integrity.[37]

Structural Discourses of Moral Injury

The juridical discourse, then, draws attention away from the commissioner-as-subject of clinical moral injury and toward an understanding of the morally injured subject as being a victim of another's actions. Although both are certainly attentive to other levels of interaction, Hampton and

---

[36] Bernstein, *Torture and Dignity*, 1; "Suffering Injustice," 305.

[37] Indeed, Bernstein goes to argue that the liberal rule of law was developed during the Enlightenment to protect people from torture, and that this foundation, now forgotten, that liberal society protects us not as inherently sacred or dignified people but as beings that are constantly at risk of losing hard-won dignity, should be put at the center of rationales for such discourses as human rights.

Bernstein focus their examples of moral injury on interpersonal violence, where moral injury arises from action taken by one individual or another. Hampton writes of postbellum black laborers being humiliated and brutally murdered by white farmers in the United States, while Bernstein focuses on torture and rape. Although the social structure is implicitly important in each case, the emphasis remains on moral injury as the result of interpersonal engagement. This is largely to do with the focus of their work. Hampton is concerned with practical jurisprudence, while Bernstein is engaged in re-establishing the foundations of the need for the liberal rule of law philosophically on the shared human vulnerability to devastation.[38] This allows this discourse to develop concepts such as moral injury and dignity to illustrate the moral ramifications of violence committed against another. At the same time, there are dimensions—the institutional, the structural—that may not be foregrounded in their discussion of specific examples of violence.

Psychologist Carol Gilligan inaugurates a discourse that, in comparison, focuses on social structure and moral development as the location of moral injury. Gilligan draws on Shay's definition, yet extends it beyond the battlefield to understand moral injury more generally as a loss of trust that affects our ability to care for others.[39] Specifically, Gilligan argues that the ways young men and women, and particularly children, are engendered and disciplined into gendered norms can stunt their ability to care for others. For women, this can come from a dilemma of either maintaining peer relationships by acquiescing to peer pressure or developing an authentic voice that, although articulating one's feelings and needs, makes one stand out socially, thus risking peer isolation. For men, this blockage can originate in the felt need to sacrifice one's feelings and self-expression so that one can fit into a socially acceptable image of manhood.[40] The social pressures males feel as males and females as females can thus inhibit theirs own more authentic development—what Gilligan calls "having a voice"—and

---

[38] Hampton's focus is juridical institutions, and so even though she is clearly aware of race, her focus on specific institutions restricts her focus, in that the goals of judicial procedure deal with groups or individuals and not with entire social structures. Bernstein's larger project is to reframe moral philosophy and to re-establish the ground of liberal rights not on notions of dignity founded neither on rationality nor on natural law or utility, but instead on what he argues is the vulnerability all humans in the face of devastation.

[39] Gilligan, "Moral Injury and the Ethic of Care," 90.

[40] Gilligan, "Moral Injury and the Ethic of Care," 92, 94.

by extension limit the development of orientations of care that allow them to relate more authentically to others.[41]

Unlike other moral injury discourses, Gilligan's structural focus does not identify a commissioner of harm. Instead, everyone in society is a subject of moral injury, and the dynamics of society and culture themselves give rise to the harm. What is critical is that, although Gilligan, like Shay, Bernstein, and Hampton, also sees the moral injury as injurious to trust, she does not see this as the result of discrete acts of violence. Instead, and more fundamentally, moral injury is part of moral development itself. The violence experienced is thus structural in that it comes from the ways in which society creates an individual's subjectivity through gendered social norms and practices. Gilligan's subject, then, is at once individual, as she speaks about individual moral development, and yet also corporate. Her subject is at once intimate and particular, social yet also more abstract. Everyone, then, participates in both the commission and subjectivity of moral injury, as everyone both internalizes and polices gender norms.

While Gilligan focuses specifically on the construction of gender, moral injury could also be used to describe the effects of other nomothetic structures, such as race. Activist, writer, and trainer Jessica Vazquez Torres has begun using the term to denote the ways in which empathy can be limited through the cultural-moral developmental process of racialization. Drawing on her own experience of working with groups in the United States to unpack the dynamics of race, Vazquez Torres argues that the experience of race can limit one's empathy for other groups and individuals who should be seen as allies in social justice work. Instead of an African-American man seeing a Puerto Rican queer woman as an ally, for example, he can see her as competition in the struggle for social recognition and economic success. Racial structures and the experience of structural violence throughout one's life can limit the imaginative, empathetic leap necessary to connect one's own struggles with those of others. As Vazquez Torres writes, racism has the power to "destroy the possibility of solidarity between minoritized groups, rendering us ineffective in our capacity to leverage our power to change the organizational structures and institutional cultures that dehumanize us."[42] Groups that should be working together to fight injustice instead attack each other, and potential white allies withdraw or assume postures that they think will make them look

---

[41] Gilligan, "Moral Injury and the Ethic of Care," 92.
[42] Vazquez Torres, "Does Moral Injury Have a Color?," 3.

non-racist—even as such moves deny some groups power while allowing others, such as whites, to avoid taking ownership of the social roles in which whiteness casts them.

The structural dynamics at play not only limit effective coalition building, however, Vazquez Torres emphasizes that they also limit moral imagination more generally. Drawing on critical race theory and other liberative discourses, Vazquez Torres argues that too often struggles about race are seen exclusively as a struggle between black and white, which elides the experiences and struggles of other communities of color. This can limit one's moral imagination and perception of who gets to be included in the high-stakes social debates about race and who is available to create coalitions for social justice.

In this way, racism can stunt the growth of empathy, as Gilligan also points out. The difference, however, is that instead of looking only at gender, Vazquez Torres opens up a structural understanding of moral injury to other structures that socialize individuals, as intersectionality is central to her understanding of moral injury and the creation of morally injurious subjectivities. At the same time, Vazquez Torres's analysis also resonates with that of Hampton and Bernstein. Racism is inherently dehumanizing, and so is an issue of dignity. This is dignity, however, understood through empathy and connected to care. If seen as a malformation of moral imagination, moral injury limits empathy because it limits the range of those who can be considered to be fully human, and so the subjects of empathy.[43] In this way, Vazquez Torres can connect Gilligan's use of care with Bernstein and Hampton's language of dignity, as Vazquez Torres sees racism as structures affecting one's empathic capacity to see a wider range of individuals as worthy of dignity and so sharing in one's own value. Vazquez Torres's usage gestures toward if not a common suffering then one related between members of a society around white supremacy, which affects everyone's moral development, even if in diverse ways.

MORAL INJURY AND MORAL SUBJECTIVITY

The origins of these different discourses show that, for the most part, they not only derive "moral injury" from different intellectual sources but also refer to different experiences and institutional locations. They understand differently the subject of moral injury, that subject's relationship to the

---

[43] Vazquez Torres, "Does Moral Injury Have a Color?," 7.

commission of their injury, as well as the context in which such injuries are possible. Moral injury as described in these pages draws from yet other sources, including writings on subjectivity and violence, as well as theories that use virtue language. What this comparison helps show, however, is the complexity of defining much of the experience of war and similar violence in moralistic terms. This is nothing new, of course. The Holocaust provides a sadly rich literature that illustrates how war, and genocide in particular, can make victims feel ultimately complicit in the attempt at their own survival or that of their people, whether it be the councils in Jewish ghettos in Nazi-occupied territories that helped in the selection of individuals for transportation to the camps (though often in an attempt to forestall more tragedy), or the ways in which inmates of concentration camps were made to contribute to the running of the camps. In a sense, moral injury as a form of durable domination by void reflects each of these discourses: the clinical, in that one can feel they have, through their agency, committed grievous wrongs that cannot be easily forgotten or forgiven; the juridical, in that one can feel dehumanized through treatment during the war, including conditions forced on the population of Sarajevo during the siege; and even the structural, as one can come to question the way they were brought up or socialized or even question the justness of their own culture, and how that culture could give birth to such violence. This last can arise from the dissonance that violence can create between one's deep assumptions of existence and the realities of war, raising questions such as the poet Semezdin Mehmedinović captures when he writes, "How can this happen/Here, of all places, where we are so humane?"[44]

Domination by void, then, can create, through duration or through degree, a lasting sense of injury. It is an injury to one's vision and imagination, as well as one's ability to enter into different relationships and social contexts, to inhabit different local moral worlds. This metaphor of injury can be misleading, however, as it refers to a physical wound, a location one can see and touch. There may be neurological correlates to moral injury, but the emphasis in describing an experience as a moral injury is to point to the harmful consequences of violence and to make more visible the suffering of beings whose existence includes a dimension of interiority. Violence is injurious to ways of seeing, ways of imagining, and ways of relating, pointing to a disruption in one's ability to live reasonably well in everyday life.

---

[44] Quoted in Lieblich and Boškailo, *Wounded I Am More Awake*, 16.

In a way, although individuals feel such injury, if there is a physical correlate, it might be more helpful to see the observable injury in the daily routines, institutions, and practices that form the ground of one's moral subjectivity. These local moral worlds, as Kleinman argues, are the sites where value and meaning are created. They have shared spaces, institutions, and practices that empower one to see daily actions as having some purpose. The violence of wars such as those in Yugoslavia during the 1990s targeted not only individuals but also communities, cultures, and the basis of identity, eliminating the possibility of certain local moral worlds. Without access to such places, without context, as Veena Das might describe it, one also loses access to those images, narratives, and norms that allow one to imagine oneself as moral, that is, striving toward virtue. It limits the conditions for the possibility of certain moral intersubjectivities, of the possibility of even imagining certain relationships and futures. It is an attack on the material and social foundation of moral selfhood and identity.

One's life, then, can be in danger not only of meaninglessness but meaninglessness defined in moral terms, as the inability to strive toward certain images of what one should be, and away from other images. I am speaking here about a key aspect of identity, and to be dominated by the void to the point of injury also harms one's ability to have a morally significant identity. This is important, as imagination is, as philosopher Victoria McGeer claims in her writing on hope, central to the agency, what we see ourselves able to do and not to do.[45] For McGeer, the consequences of looking at the agency in this way are that how we see ourselves as agents, and how we imagine ourselves relating to others, are central to identity. Transformations to the ability to imagine are also, then, transformations of identity. Such imagination is moral, then, not because there is a separate faculty from imagination called the *moral* imagination. Instead, the imagination is itself moral because it is central to identity and identity formation, to the way we see ourselves developing as good or bad persons, how we should relate to and treat others, as well as the ways in which we can imagine different futures, as well as paths out of conflict.[46] It may even

---

[45] McGeer, "The Art of Hope," 104. This section on McGeer is shared with an article, "'A Distress that Cannot Be Forgotten:' Imagination, Injury, and Moral Vulnerability," forthcoming in *Philosophy Today*.

[46] There is more to be said here, as I begin to enter the realm of the "social imaginary," as Charles Taylor has framed it, of the possible images and narratives available to us given our time and cultural space. Violence certainly affects such an imaginary, which then affects our

be accurate to say that one does inhabit a world here but it is an impoverished one, where the world is vicious, or oneself is vicious, or where the possibility of meaning and goodness is lost.

Murdoch writes of the void, toward the end of *Metaphysics as a Guide to Morals,* that the world we live in may, in fact, be filled with more miserable, hopeless persons than the obverse.[47] To be morally injured is to be made painfully aware of the vulnerability of living in such a world. The very discourse of injury assumes vulnerability, and in this case, it is a vulnerability in one's moral subjectivity and identity, resting as they do on practices, norms, institutions, and the actions of others, all of which are either beyond one's control or are resistant to one's influence. It is a vulnerability, a moral vulnerability, inherent in the human understood as a species that needs to be able to strive toward the *Good* that needs to feel it is a possibility in the world, and where striving for goodness is part of one's identity. There is, to speak in Bernstein's speech, a breach of trust with the world that goodness is possible and that the world is of such a nature that it nurtures goodness. If the world's nature is not so, if one's own nature is also not so, then one's identity has changed to one where he or she is no longer good or capable of goodness. With no good identity to turn to, one can become lost, and even despair dislocated and in search of images and concepts through which they can create new moral lives. The result can be an injury, a moral injury, a morally injured subjectivity that alters the very possibility of what one can imagine.

ability to imagine, form identity, and create and sustain community. Indeed, the way in which violence arises from everyday life and then is folded into such life is a key aspect of Das's and Kleinman's work.

[47] Murdoch raises this question, but distances herself from such strong claims. She writes, "The average inhabitant of the planet is probably without hope and starving. It is terrible to be human" (Murdoch, *Metaphysics as a Guide to Morals,* 498). She begins this particular paragraph, however, with the phrase, "Someone may say." By doing so, it is unclear whether Murdoch is presenting a particular argument, which seems likely, or whether she agrees with this argument. She does this throughout her discussion of void, as if she is dealing with issues that are difficult to discuss head-on.

CHAPTER 7

Conclusion

This work began with research into the intersection of religion, violence, and ethics. Over several years I became curious, as a religious and social ethicist, about the lack of vocabulary to describe such experiences adequately. My curiosity became more acute as I reflected on the vocabularies I was already aware of in moral philosophy and religious ethics, particularly vocabularies of virtue that are so rich in their descriptive quality and representation of the human as a moral being. Such resources were not to be found in the social sciences, at least not significantly. This is changing, and ethics and virtue are now legitimate areas of inquiry. In the study of political violence, however, there still seems an inability, even when intentionally probing the moral dimensions of violence, to put the experience of moral transformation at the heart of our inquiries. This no doubt arises from the fact that social-scientific disciplines that have political violence as one of their issues come from a history where the ethical, understood as part of the theological, was one domain against which their fields were defined in the nineteenth century. History being history, origins are not so easily overcome. The experience of some individuals in the Bosnian War, as well as in other conflicts, may nevertheless be a moment where this legacy of the nineteenth century can perhaps be addressed elaborately. Work done by anthropologists and sociologists mentioned in these pages points to bridges being built.

There remains, however, one-sided nature of this endeavor. The discussion about virtue and ethics seems to be taking place largely within

© The Author(s) 2019
J. Wiinikka-Lydon, *Moral Injury and the Promise of Virtue*,
https://doi.org/10.1007/978-3-030-32934-1_7

disciplines and not across disciplines. The thought partners for these for-
ays into the resources of ethics seem to be coming from within one's own
discipline. This is a broad generalization, yet it may point to the work still
needing to be done. Using the resources of another tradition and discipline without engaging with those working in those traditions would only
reaffirm disciplinary boundaries and may not lead to many transformations. Moral philosophy still has a great deal to learn from the social sciences, as do religious ethics, though religious social ethics has for over a
century been deeply engaged with the social sciences.[1]

This book is an attempt to show how work from the humanities can
engage with that from the social sciences to create greater insight into
issues, such as political violence, that are important not just to different
disciplines but to society as a whole. I am not going to argue here for
interdisciplinarity or even transdisciplinarity, as that has been done elsewhere, in more detail, and more skillfully than I can do here. Instead, this
work should serve as an example of one way in which this can be done, and
how engagement not of ethics but by ethicists can produce different
insights into current subjects in the social sciences. Likewise, it should
serve as an example of how resources in moral philosophy and religious
ethics can be heightened by bringing in the insights of other disciplines.
The engagement of Murdoch's moral subjectivity with Kleinman's understanding of local moral worlds is an example of this, that I hope to expand
even more in future work.

Equally important is the humanistic commitment that I hope animates
these pages, to understand social and political suffering and, through better understanding, to help mitigate or even ameliorate it. Any academic
work must be deeply modest about its contribution to such work, but
nevertheless, there are contributions to be made. The historian of religion
Jonathan Z. Smith saw a responsibility on the part of the humanistic sciences to make intelligible events that seem unintelligible. Smith's occasion was the deaths at Jonestown, Guyana, in November 1978, but there
are sadly many more from which we could draw.[2] These are the events
that in their immediate aftermath individuals often call "unimaginable,"

---

[1] For the history of Christian social ethics, see Dorrien, *Social Ethics in the Making*. For
recent arguments about religious ethics turning toward anthropology or critical cultural
studies, see Banner, *The Ethics of Everyday Life*; Miller, *Friends and Other Strangers*; Scharen
and Vigen, *Ethnography as Christian Theology and Ethics*.

[2] Smith, *Imagining Religion*, 120.

"irrational," and "unthinkable." That human beings plan such violence undermines claims of unthinkability, yet perhaps the fact that we do claim that certain acts are unthinkable can be seen less as a claim of what is thinkable and more a prompt, even a plea, to help one think through horror and the urgency horror can pose in making room for its possibility in one's imagination and view of the world. Helping one to think better about events and dynamics that seem to inherently counter to sense is an important social role for the academic, and not only for those in the social sciences. Indeed, as we are speaking of human events and needs, all disciplines are needed in attempts to expand our capacity to speak about matters of dire import.

Where there is no language to make this move from shock and stasis to thoughtful action, the scholarship can be of use here, bringing not only analysis, but also—prior to this and in support of analysis and critique—languages, sociologies, and social-scientifically informed philosophical anthropologies. Such work can help the shocked voice to speak again, at least, if individuals find such resources helpful. That is a critical point. Such resources must always be seen as offerings to those, both in academia and without, working and even struggling to articulate and make sense of world-altering events.

What struck me in researching this was not just the moral stakes involved but the difficulty in giving voice to it. This was partly to do with the seeming absurdity of what people experienced. Only the day before citizens of a cosmopolitan European city, Sarajevans experienced a surreal, even tragicomic expulsion from that reality. In one grim episode, underscoring the unwelcome irony of the siege, a busy market was bombed in Sarajevo on the tenth anniversary of the 1984 Sarajevo Winter Olympics. As the bodies were buried in an Olympic practice field next to an Olympic hall, the opening ceremonies of the latest games were being broadcast live from Lillehammer, Norway. For a city that considered itself European and cosmopolitan, embodied in the pride of being an Olympic city, such an irony became a symbol of lost pride, the loss of those characteristics that had made the city what it was. And for the Olympics to go on, even as an Olympic city was being starved and besieged, seemed to slash and scar the multicultural, global ethics, and subjectivities that those games would seem to embody.

The difficulty in articulating the experience and deep subjective repercussions of violence should not, however, really be surprising. How many of us are raised with a vocabulary built to attend to days of horror? Even with the violence that abounds in film and games, there is at the same time the lack of a vocabulary to allow us to articulate the moral loss that can come with violence. Why this is so must wait for further work, but that this is important to acknowledge. Indeed, I took my cue at the beginning of this work from Zlatko Dizdarević and his reflections during the Bosnian War, as well as from Maček's ethnography. As a journalist on Sarajevo's leading newspaper, Dizdarević already had a vocation to try and give voice to what was happening in his city and country. Throughout the diary, the reader can see this struggle. At times, civilization and the culture of Sarajevo are doomed. In others, it is the attacker that is doomed, and those within the city, the besieged, who have managed to hold on to their humanity in spite of the horrors. I take this not as confusion but as some-one wrestling with the immediate experience of the wanton destruction of his home and his need to communicate his experience to others and even to make it intelligible to himself.

We also saw this in Maček's work, which describes how her informants tried to make sense of the war, and yet how just as often they would fail in their attempts.[3] Maček's informants themselves were struggling to find the words to fully articulate the stakes involved in what they were sensing as a loss of something central to who they were. Key here is Maček's account of the concern for *normality*. Individuals did not have a vocabulary to fully explicate this concern, but they were nevertheless motivated and even driven by emotions such as shame, worry, fear, and maybe resentment, all communicating that they had fallen from something called "normality." Something important was lost that needed to be recovered, and much of life during the siege, as Maček recounts it, was taken up with activity related to being "normal."

This normality is essentially a word to refer to an idealized world that was lost, as well as the anxiety that one no longer was able to live in that world. Normal was life in pre-conflict Bosnia with its routines, standards of dress, and behavior, but also importantly a life where one felt that they could live as a basically good person. It was, in other words, a world that made sense on multiple levels. The feeling that one is no longer normal or a good person is felt deeply, yet it is not necessarily something that can be put easily into words. One can experience it more as a sensation, a mood, a loss of motivation or trajectory, a location, an inhabitation, a haunting. The survivor can continue with life, unable to express to someone outside

---

[3] Maček, *Sarajevo Under Siege*, 4, 6.

that experience that a profound change has occurred. What they do have are the bodily feelings and cognitively tied emotions that make one's self and one's world feel quite different than before. It is largely the emotions that tell one something is wrong, even if such knowledge needs to be interpreted further.

In trying to respond to these witnesses there are very powerful resources in the social sciences. This, however, also poses a challenge, as the power of such disciplines and their analysis can threaten to subsume moral experience within or reduce it to discourses of economics, politics, history, ethnicity, or any other category. It is too easy to elide moral experience in discussions of social dynamics, even when the purpose of one's study is to discuss the moral dimension of local social dynamics. And it is too easy to treat the moral experience as epiphenomenal and reduce it to biological, cognitive, or physiological substrates. What is needed are ways to frame the discussion that tends toward understandings of individuals as first and foremost moral subjects whose negotiation through existence is morally significant, both to them and to their communities and networks. Morally significant, yet also we are moral subjects whose very subjectivity is one of change, and so we need ways to articulate and account for the experience of being morally transformed. We need ways to make such transformation and this significance central not only to our research but also to the assumptions that make up the methodologies that shape that research.

I have argued, then, for a broad frame that I have called "*moral subjectivity*" to give a focus and foundation for a virtue understanding of the self. This conception of moral subjectivity includes certain characteristics, such as worldview, ethos, practices, and so on. What is most important, however, is that framing experience in terms of moral subjectivity opens one's research to engagement with moral vocabularies and theories. It is a frame that emphasizes the conditioned nature of the human self. Institutions, structures, cultural narratives, symbols—all of this help give rise to the phenomenon called "*the self*." Grounded in basic social theory, a frame of moral subjectivity is open to moral theories and theories of moral development and moral phenomenology. This creates a frame that accounts for the embedded sociality of subjectivity but also helps to articulate the experience of being such an embedded, embodied subject. We are able, then, to see how changes to local institutions during the Bosnian War, for example, shaped one's felt ability to be a "good" person, while also being able to account for what aspects of one's subjectivity were transformed and to articulate how that was experienced using morally laden terminology. What we end up with is what I have called a *"virtue herme-neutic"* meant to give voice to certain experiences where vocabulary is lacking, both for those experiencing violence and for those who study it. It is, then, an interpretation of that experience, an aid to understanding what has happened.

Throughout this work, my emphasis has been on the experience of extreme violence or extreme situations. I take a multi-year siege, as well as genocide and crimes against humanity, to be extreme, and where many years are usually required to alter the trajectory of one's moral develop-ment, such extreme situations can compress that time, just as heat and pressure do to carbon to create coal, and create felt changes in one's moral subjectivity in very short periods. By using the word "extreme" I do not mean rare. We live in a world where extreme violence is, if not prevalent, then not uncommon, a fact that can desensitize us to the way in which such violence can still have extremely injurious consequences.

Perhaps due to the extreme nature of the times, such changes can be lasting. The doggedness of certain types of moral injury and harm can attest to this. Much too often, I would argue, ethicists, engage in thought experiments that use extreme, rare, particular examples to make univer-sally applicable arguments. This methodology is dubious, not the least of which because such experiments are rarely grounded in actual cases nor in the particulars that such particularity of the thought experiment would seem to promise. Political violence is not uncommon, although the expe-rience of being besieged is rarer. The practices of domination and oppres-sion that occur in such sieges, however, are sadly not rare, as those practices, including rape, racism, torture, strategic neglect, and so on., are found throughout society.

The hope here is that work in the future can build on the goals of the present study and see how virtue can be used to help articulate the experi-ence of violence such as racism, sexism, and others. The only claim I can make in that regard is that virtue discourse is largely untapped for such practical endeavors, even with centuries of study. It is also, then, *richly* untapped and can provide ways of articulating experiences of society, using commonly held vocabulary, to help illustrate to others who do not experi-ence such violence the long-lasting and profoundly deep consequences of violence and social injustice. In this way, this is a study that addresses cer-tain contexts of violence as a way to contribute more broadly, through a specified inquiry, to the constant problem of political and social violence.

This book was also methodological in focus, yet methodology is perhaps one of the most ethically charged locations in one's scholarship. It is here that assumptions and approaches are articulated or hidden, where culture and social location perhaps first permeate inquiry. It is also here that the general approach of those like Iris Murdoch can contribute even more, and perhaps most profoundly, to social-scientific methodology and inquiry. Murdoch, and possibly other thinkers engaging virtue discourse and philosophical anthropologies and metaphysics grounded in virtue, provides a way to describe this view of being a human being. It is a view, as I have argued, that is not only local, as Kleinman argues, but also eye level. Murdoch brings us down to the individual senses and stresses what the human senses. This is not the researcher's abstracted view looking down on society, but rather a view from the ground, so to speak. At least, that is its potential. Murdoch conceives of an image of the human as a moral subject for whom something as basic as sight (or sensation and perception more generally) is shot through with significance. What, and especially how, one sees matters morally to a great and profoundly fundamental degree.

When applied to political violence, such an image of the human draws one's analysis to eye level and emphasizes what happens at that level of interaction. What does one see? How are the conditions of moral perception affected? And Murdoch grounds this further in an understanding of the *Good* to underscore how our imagination and perception are engaging not just with the moral fictions of our minds but also in something we understand to be real. She does not offer a great deal of detail about what this *Good* consists of, yet that is not critical, at least not for the purposes of this project. What is important is to emphasize that when engaging those who say they can no longer be good, we examine this as a real claim about real experiential disorientation. The experience of not being able to be good is not a fantasy. Instead, it is a reaction to a very real experience and changes in conditions that enable moral subjectivity. Whether in the end one adopts a high understanding of the *Good* as a metaphysical domain, or sees the *Good* through a low understanding as an affirmation of the way that humans understand themselves and their lives through engagement with others and the world, positing the *Good* is a move to affirm the moral life as real and not as merely epiphenomenal to the reality of the economy, the state, society, and so on.

Methodology, then, is not just an aspect of scientific inquiry but a deeply human practice. We all engage throughout our day in presupposed methodologies enabling us to interpret or engage every day. In an academic context, this process is more reflexive and formalized but it still contains a ubiquitous human practice in which we situate ourselves and through which we understand our lives. It is, then, an aspect of research wherein responsibility is centrally relevant. Methodologies arise from a response to the world, as some issue or experience spurs us to respond to it in an academic mode. How we respond, and what frames we use in that response, will shape the conclusions we come to about society and individuals (in other words, our neighbors and communities).

The human sciences cannot be intelligible without rich accounts of the human that prioritize moral development and approach social life as moral life. This does not mean that anthropological or sociological inquiries need to take firm prescriptive stances, nor does it mean that the methods of such inquiries need to be based on a form of moral or moralizing authority. Instead, for research to make primary the moral dimension of social life will better reflect the ways in which individuals and communities understand and practice their own lives. Indeed, such a stance reflects the way that researchers view their own life when away from their professional work.

It is here, then, that methodology can be seen as a location, a practice, a hope. This is an important way of seeing research, particularly research that deals with people in desperate circumstances, where not only their livelihoods but also their souls are targeted. Giving people a way to articulate what is at stake in the experience of violence, domination, and oppression in terms used by the broader population provides a beginning to make the claims of survivors more intelligible to those of us for whom such violence is alien and for whom claims of violence can be suspect. There is, then, hope implicit in unheard voices speaking, even if only to acknowledge how quickly and violently our very souls can be transformed. Methodology can, then, be a practice of hope, and as any such practice, one that creates virtues not only in the researcher but possibly in the reader and listener as well. Such practices that help make scholarship truly interdisciplinary but also help create bridges between different domains of society will by the necessity of our finitude stand to exceed our expectations by leaving what we can control and affecting broader society.

Perhaps such a language is too optimistic. But, hopefully, such work can provide avenues for the humanistic task of relieving sufferings by putting whatever resources and wisdom we have earned in the hands of those who have experienced society's violence. Humanistic writing and research, in dialogue and engagement with other fields, is a potential meeting place where different individuals and communities can attend to the injuries suffered from war and violence. Often seen as formal and dry, even unemotional, activities such as research, methodology, and analysis can also be places that generate care and assistance. Research can reach out and respond to the sufferings and needs of others, making it a truly humanistic work. And, even the attempt to try and recognize another's suffering can itself create new insights, as well as new relationships and ways of being philosophers, ethicists, theologians, social scientists, that are humbly reflective of the imaginative transformation so often needed for the moral repair of injurious subjectivities. After a whole work engaging extreme violence and genocide, perhaps a little hope and even sentimentality is earned. For these reasons, I turn to the poets to have the final, hopefully, say on humanistic methodology:

> Take the emptiness you hold in your arms
> and scatter it into the open spaces we breathe:
> maybe the birds will feel how the air is thinner,
> and fly with more affection.[4]

---

[4] Rilke, *Duino Elegies*, 10.

BIBLIOGRAPHY

Ali, Rabia, and Lawrence Lifschultz, eds. *Why Bosnia? Writings on the Balkan War.* Stony Creek, Conn: Pamphleteers Press, 1994.

Allen, Beverly. *Rape Warfare: The Hidden Genocide in Bosnia-Herzegovina and Croatia.* Minneapolis: University of Minnesota Press, 1996.

Annas, Julia. "Virtue and Eudaimonism." *Social Philosophy and Policy* 15:1 (December 1998): 37–55.

———. *Intelligent Virtue.* New York: Oxford University Press, 2011.

Anscombe, G.E.M. "Modern Moral Philosophy." *Philosophy* 33:124 (1958): 1–19.

Antonaccio, Maria. "Moral Change and the Magnetism of the Good." *Annual of the Society of Christian Ethics* 20 (2000a): 143–64.

———. *Picturing the Human: The Moral Thought of Iris Murdoch.* New York: Oxford University Press, 2000b.

———. "The Virtues of Metaphysics: A Review of Iris Murdoch's Philosophical Writings." In *Iris Murdoch, Philosopher*, edited by Justin Broackes, 155–79. New York: Oxford University Press, 2009.

———. *A Philosophy to Live by: Engaging Iris Murdoch.* New York: Oxford University Press, 2012.

———. "Symposium on Iris Murdoch: A Response to Nora Hämäläinen and David Robjant." *The Heythrop Journal* 54:6 (November 2013): 1012–20.

Aristotle. *The Politics.* Translated by Carnes Lord. Chicago: University of Chicago Press, 1985.

Banac, Ivo. *The National Question in Yugoslavia: Origins, History, Politics.* Ithaca, NY: Cornell University Press, 1988.

Banner, Michael. *Ethics of Everyday Life: Moral Theology, Social Anthropology, and the Imagination of the Human.* New York: Oxford University Press, 2016.

© The Author(s) 2019
J. Wiinikka-Lydon, *Moral Injury and the Promise of Virtue*,
https://doi.org/10.1007/978-3-030-32934-1

Bellah, Robert N., Richard Madsen, William M. Sullivan, Ann Swidler, and Steven M. Tipton. *Habits of the Heart: Individualism and Commitment in American Life*. Berkeley: University of California Press, 2007.

Benimoff, Roger, and Eve Conant. *Faith Under Fire: An Army Chaplain's Memoir*. New York: Broadway Books, 2010.

Bernstein, J.M. "Suffering Injustice: Misrecognition as Moral Injury in Critical Theory." *International Journal of Philosophical Studies* 13 (2005): 303–24.

———. "Suffering Injustice: Misrecognition as Moral Injury in Critical Theory." *International Journal of Philosophical Studies* 13 2015. *Torture and Dignity: An Essay on Moral Injury*. Chicago: University of Chicago Press.

Berkwitz, Stephen C. "History and Gratitude in Theravada Buddhism." *Journal of the American Academy of Religion* 71:3 (September 2003): 579–604.

Besirevic-Regan, Jasmina. "The Ethnic Cleansing of Banja Luka: National Homogenization, Political Repression, and the Emergence of a Bosnian Muslim Refugee Community." Doctoral dissertation, Yale University, 2004.

Biehl, Joao, Byron Good, and Arthur Kleinman. *Subjectivity: Ethnographic Investigations*. Berkeley: University of California Press, 2007.

Blum, Lawrence. "Visual Metaphors in Murdoch's Moral Philosophy." In *Iris Murdoch, Philosopher*, edited by Justin Broackes, 307–24. New York: Oxford University Press, 2014.

Boudreau, Tyler. "The Morally Injured." *The Massachusetts Review* 52 (2011): 746–54.

Bourdieu, Pierre. *Outline of a Theory of Practice*. Translated by Richard Nice. New York: Cambridge University Press, 1977.

———. *Pascalian Meditations*. Translated by Richard Nice. Stanford, Calif: Stanford University Press, 2000a.

———. *The Weight of the World: Social Suffering in Contemporary Society*. Translated by Priscilla Parkhurst Ferguson. Stanford, Calif: Stanford University Press, 2000b.

Bourgois, Philippe. "The Power of Violence in War and Peace: Post-Cold War Lessons from El Salvador." *Ethnography* 2:1 (2001): 5–34.

Bringa, Tone. *Being Muslim the Bosnian Way*. Princeton, N.J.: Princeton University Press, 1995.

Broackes, Justin. "Introduction." In *Iris Murdoch, Philosopher*, edited by Justin Broackes, 1–92. New York: Oxford University Press, 2014a.

———, ed. *Iris Murdoch, Philosopher*. New York: Oxford University Press, 2014b.

Brock, Rita Nakashima, and Gabriella Lettini. *Soul Repair: Recovering from Moral Injury after War*. Boston: Beacon Press, 2013.

Brubaker, Rogers. *Nationalism Reframed: Nationhood and the National Question in the New Europe*. New York: Cambridge University Press, 1996.

———. *Ethnicity Without Groups*. Cambridge, Mass: Harvard University Press, 2006.

Bryan, Anna Belle O., Craig J. Bryan, Chad E. Morrow, Neysa Etienne, and Bobbie Ray-Sannerud. "Moral Injury, Suicidal Ideation, and Suicide Attempts in a Military Sample." Traumatology 20 (2014): 154–60.

Campbell, David. "Violence, Justice, and Identity in the Bosnian Conflict." In *Sovereignty and Subjectivity*, edited by Jenny Edkins, Nalini Persram, and Veronique Pin-Fat, 2–38. Boulder, CO: Lynne Rienner, 1999.

Cigar, Norman. *Genocide in Bosnia: The Policy of "Ethnic Cleansing"*. College Station, TX: Texas A&M University Press, 1995.

Cohen, Leonard J. "Bosnia's 'Tribal Gods': The Role of Religion in Nationalist Politics." In *Religion and the War in Bosnia*, edited by Paul Mojzes, 43–73. Atlanta: Scholars Press, 1998.

Conradi, Peter J. "*Divine Though Unfinished*: Letters to Roly Cochrane." *The Iris Murdoch Newsletter* 19 (2006), 29–30.

Copp, David, and David Sobel. "Morality and Virtue: An Assessment of Some Recent Work in Virtue Ethics." *Ethics* 114 (April 2004): 514–54.

Crisp, Roger and Michael Slote, eds. *Virtue Ethics*. Oxford: Oxford University Press, 1997.

Crocq, Marc-Antoine, and Louis Crocq. "From Shell Shock and War Neurosis to Post-traumatic Stress Disorder: A History of Psychotraumatology." *Dialogues in Clinical Neuroscience* 2:1 (March 2000): 47–55.

Cuk, Nadia Skenderovic. "Temptations of Transition and Identity Crisis in Post-Communist Countries: The Example of Former Yugoslavia." In *Ethnic Conflict and Civil Society: Proposals for a New Era in Eastern Europe*, edited by Andreas Klinke, Ortwin Renn, and Jean-Paul Lehners, 83–98. Brookfield, VT: Ashgate, 1998.

Das, Veena. "The Act of Witnessing: Violence, Poisonous Knowledge, and Subjectivity." In *Violence and Subjectivity*, edited by Veena Das, Arthur Kleinman, Mamphela Ramphele, and Pamela Reynolds, 205–25. Berkeley: University of California Press, 2000.

———. *Life and Words: Violence and the Descent into the Ordinary*. Berkeley: University of California Press, 2007.

———. "Ordinary Ethics." In *A Companion to Moral Anthropology*, edited by Didier Fassin, 133–49. Malden, MA: Wiley-Blackwell, 2012.

Das, Veena, Arthur Kleinman, Margaret M. Lock, Mamphela Ramphele, and Pamela Reynolds, eds. *Remaking a World: Violence, Social Suffering, and Recovery*. Berkeley: University of California Press, 2001.

Das, Veena, Arthur Kleinman, Mamphela Ramphele, and Pamela Reynolds, eds. *Violence and Subjectivity*. Berkeley: University of California Press, 2000a.

Demick, Barbara. *Logavina Street: Life and Death in a Sarajevo Neighborhood*. New York: Spiegel & Grau, 2012.

Denitch, Bogdan. *Ethnic Nationalism: The Tragic Death of Yugoslavia*. Minneapolis: University of Minnesota Press, 1996.

Dizdarević, Zlatko. *Sarajevo: A War Journal*. New York: Fromm International, 1993.

Drescher, Kent D., David W. Foy, Caroline Kelly, Anna Leshner, Kerrie Schutz, and Brett Litz. "An Exploration of the Viability and Usefulness of the Construct of Moral Injury in War Veterans." *Traumatology* 17 (2011): 8–13.

Driver, Julia. *Uneasy Virtue*. New York: Cambridge University Press, 2007.

Eagleton, Terry. *Figures of Dissent: Critical Essays on Fish, Spivak, Žižek and Others*. New York: Verso, 2005.

Farley, Margaret A. "The Role of Experience in Moral Discernment." In *Christian Ethics: Problems and Prospects*, edited by Lisa Sowle Cahill and James F. Childress, 134–51. Cleveland: Pilgrim Press, 1996.

Fassin, Didier. *A Companion to Moral Anthropology*. Malden, Mass: Wiley-Blackwell, 2012b.

———. "Introduction." In *A Companion to Moral Anthropology*, 1–18. Malden, Mass: Wiley-Blackwell, 2012.

Flanagan, Kieran, and Peter C. Jupp, eds. *Virtue Ethics and Sociology: Issues of Modernity and Religion*. New York: Palgrave Macmillan, 2001.

Foucault, Michel. *Discipline and Punish: The Birth of the Prison*. Translated by Alan Sheridan. New York: Vintage Books, 1995.

Frankl, Viktor E. *Man's Search for Meaning*. Boston: Beacon Press, 2006.

Fricker, Miranda. *Epistemic Injustice: Power and the Ethics of Knowing*. New York: Oxford University Press, 2007.

Galtung, Johan. "Violence, Peace, and Peace Research." *Journal of Peace Research* 6:3 (January 1969): 167–91.

Geertz, Clifford. *The Interpretation of Cultures*. New York: Basic Books, 1977.

Gilligan, Carol. "Moral Injury and the Ethic of Care: Reframing the Conversation about Differences." *Journal of Social Philosophy* 45:1 (2014): 89–106.

Gilligan, James. "Shame, Guilt and Violence." *Social Research* 70:4 (Winter 2003): 1149–80.

Glenny, Misha. *The Fall of Yugoslavia: The Third Balkan War*. New York: Penguin Books, 1996.

Gorski, Philip. "Recovered Goods: Durkheimian Sociology as Virtue Ethics." In *The Post-Secular in Question: Religion in Contemporary Society*, edited by Philip Gorski, David Kyuman Kim, John Torpey, and Jonathan Van Antwerpen, 77–104. New York: New York University Press, 2012.

———, ed. *Bourdieu and Historical Analysis*. Durham, NC: Duke University Press, 2013.

Hall, Pamela M. "The Mysteriousness of the Good: Iris Murdoch and Virtue Ethics." *American Catholic Philosophical Quarterly* 64:3 (Summer 1990): 313–29.

———. "Limits of the Story: Tragedy in Recent Virtue Ethics." *Studies in Christian Ethics* 17 (December 2004): 1–10.

Hampton, Jean. 1991. "A New Theory of Retribution." In *Liability and Responsibility: Essays in Law and Morals*, edited by R.G. Frey and Christopher W. Morris, 377–414. New York: Cambridge University Press.

———. "Correcting Harms versus Righting Wrongs: The Goal of Retribution." *UCLA Law Review* 39 (1991–2): 1659–1702.

Heim, Maria, and Anne Monius. "Recent Work in Moral Anthropology." *Journal of Religious Ethics* 42:3 (2014): 385–92.

Herdt, Jennifer A. *Putting On Virtue: The Legacy of the Splendid Vices.* Chicago: University of Chicago Press, 2012.

Holland, Margaret. "Social Convention and Neurosis as Obstacles to Moral Freedom." In *Iris Murdoch, Philosopher*, edited by Justin Broackes, 255–73. New York: Oxford University Press, 2009.

Human Rights Watch. "'Ethnic Cleansing' Continues in Northern Bosnia. Human Rights Watch report, November 1994a. New York: Human Rights Watch.

———. "War Crimes in Bosnia-Herzegovina: U.N. Cease-Fire Won't Help Banja Luka." Human Rights Watch report, June 1994b. New York: Human Rights Watch.

Hursthouse, Rosalind. "After Hume's Justice." *Proceedings of the Aristotelian Society* 91 (1990): 229–54.

———. *On Virtue Ethics.* New York: Oxford University Press, 2001.

James, William. *The Varieties of Religious Experience: A Study in Human Nature.* New York: Penguin Classics, 1982.

Keenan, Melinda J., Vicki A. Lumley, and Robert B. Schneider. "A Group Therapy Approach to Treating Combat Posttraumatic Stress Disorder: Interpersonal Reconnection through Letter Writing." *Psychotherapy* 51 (2014): 546–54.

Kleinman, Arthur. "Pain and Resistance: The Deligitimation and Relegitimation of Local Worlds." In *Pain as Human Experience: An Anthropological Perspective*, edited by Mary-Jo DelVecchio Good, Paul E. Brodwin, Byron J. Good, and Arthur Kleinman, 169–97. Berkeley: University of California Press, 1994.

———. *Writing at the Margin: Discourse Between Anthropology and Medicine.* Berkeley: University of California Press, 1995.

———. "'Everything That Really Matters:' Social Suffering, Subjectivity, and the Remaking of Human Experience in a Disordering World." *The Harvard Theological Review* 90:3 (July 1997a): 315–35.

———. "Experience and Its Moral Modes: Culture, Human Conditions, and Disorder." In *The Tanner Lectures on Human Values*, 355–420. Salt Lake City: University of Utah Press, 1999a.

———. "Moral Experience and Ethical Reflection: Can Ethnography Reconcile Them? A Quandary for 'The New Bioethics.'" *Daedalus* 128:4 (October 1999b): 69–97.

———. "The Violences of Everyday Life: The Multiple Forms and Dynamics of Social Violence." In *Violence and Subjectivity*, edited by Veena Das, Arthur Kleinman, Mamphela Ramphele, and Pamela Reynolds, 226–41. Berkeley: University of California Press, 2000.

Kleinman, Arthur, Veena Das, and Margaret M Lock, eds. *Social Suffering.* Berkeley: University of California Press, 1997b.

Kleinman, Arthur, and Erin Fitz-Henry. "The Experiential Basis of Subjectivity: How Individuals Change in the Context of Societal Transformation." In *Subjectivity: Ethnographic Investigations*, edited by Joao Biehl, Byron Good, and Arthur Kleinman, 52–65. Berkeley: University of California Press, 2007.

Kleinman, Arthur, and Kleinman, Joan. "How Bodies Remember: Social Memory and Bodily Experience of Criticism, Resistance and Delegitimation Following China's Cultural Revolution." *New Literary History* 23:3 (Summer 1994): 707–23.

Kleinman, Arthur, Yunxiang Yan, Jing Jun, Sing Lee, Everett Zhang, Pan Tianshu, Wu Fei, and Jinhua Guo. *Deep China: The Moral Life of the Person*. Berkeley: University of California Press, 2011.

Kolstø, Pål, ed. *Media Discourse and the Yugoslav Conflicts*. Brookfield, VT: Ashgate, 2012.

Laidlaw, James. *The Subject of Virtue: An Anthropology of Ethics and Freedom*. New York: Cambridge University Press, 2013.

Lambek, Michael. "Toward an Ethics of the Act." In *Ordinary Ethics: Anthropology, Language and Action*, edited by Michael Lambek, 1–39. New York: Fordham University Press, 2010.

Laverty, Megan. *Iris Murdoch's Ethics: A Consideration of Her Romantic Vision*. New York: Bloomsbury Academic, 2007.

Levi, Primo. *The Drowned and the Saved*. New York: Vintage, 1989.

Lieblich, Julia and Esad Boškailo. *Wounded I Am More Awake: Finding Meaning after Terror*. Nashville, TN: Vanderbilt University Press, 2012.

Litz, Brett T., Leslie Lebowitz, Matt J. Gray, and William P. Nash. *Adaptive Disclosure: A New Treatment for Military Trauma, Loss, and Moral Injury*. New York: The Guilford Press, 2015.

Litz, Bret T., Nathan Stein, Eileen Delaney, Leslie Lebowitz, William P. Nash, Caroline Silva, and Shira Maguen. "Moral Injury and Moral Repair in War Veterans: A Preliminary Model and Intervention Strategy." *Clinical Psychology Review* 29 (2009): 695–706.

Luhrmann, Tanya M. "Subjectivity." *Anthropological Theory* 6:3 (September 2006): 345–61.

Maček, Ivana. *Sarajevo Under Siege: Anthropology in Wartime*. Philadelphia: University of Pennsylvania Press, 2011.

MacIntyre, Alasdair. *Dependent Rational Animals: Why Human Beings Need the Virtues*. Chicago: Open Court, 2001.

———. *After Virtue: A Study in Moral Theory*. Notre Dame, Ind: University of Notre Dame Press, 2007.

Malcolm, Noel. *Bosnia: A Short History*. New York: New York University Press, 1996.

Mansfield, Nick. *Subjectivity: Theories of the Self from Freud to Haraway*. New York: New York University Press, 2000.

Margalit, Avishai. *The Decent Society*. Translated by Naomi Goldblum. Cambridge, Mass: Harvard University Press, 1998.

Marshall, Ellen Ott. *Though the Fig Tree Does Not Blossom: Toward a Responsible Theology of Christian Hope*. Nashville, TN: Abingdon Press, 2006.

Mattingly, Cheryl. "Two Virtue Ethics and the Anthropology of Morality." *Anthropological Theory* 12:2 (2012): 161–84.

McGeer, Victoria. "The Art of Good Hope." *The Annals of the American Academy of Political and Social Sciences* 592 (March 2004): 100–127.

Mehmedinović, Semezdin. *Sarajevo Blues*. Translated by Ammiel Alcalay. San Francisco: City Lights Publishers, 2001.

Mehta, Uday Singh. *Liberalism and Empire: A Study in Nineteenth-Century British Liberal Thought*. Chicago: University of Chicago Press, 1999.

Merleau-Ponty, Maurice. *Phenomenology of Perception*. Translated by Donald Landes. New York: Routledge, 2013.

Meyers, Diana T. *Subjection and Subjectivity: Psychoanalytic Feminism and Moral Philosophy*. New York: Routledge, 2014.

Miller, Richard B. *Friends and Other Strangers: Studies in Religion, Ethics, and Culture*. New York: Columbia University Press, 2016.

Miller, Sarah Clark. "Moral Injury and Relational Harm: Analyzing Rape in Darfur." *Journal of Social Philosophy* 40 (2009): 504–523.

Mojzes, Paul, ed. *Religion and the War in Bosnia*. Atlanta: Scholars Press, 1998.

Moran, Richard. "Iris Murdoch and Existentialism." In *Iris Murdoch, Philosopher*, edited by Justin Broackes, 181–96. New York: Oxford University Press, 2014.

Mott, Frederick Walker. *War Neuroses and Shell Shock*. London: H. Frowde, 1919.

Mulhall, Stephen. "Constructing a Hall of Reflection: Perfectionist Edification in Iris Murdoch's *Metaphysics as a Guide to Morals*." *Philosophy* 72:280 (April 1997): 219–39.

———. "'All the World Must Be Religious:' Iris Murdoch's Ontological Arguments." In *Iris Murdoch: A Reassessment*, edited by Anne Rowe. London: Palgrave Macmillan, 2006.

Murdoch, Iris. *The Fire & The Sun: Why Plato Banished the Artists*. New York: Clarendon Press, 1977.

———. *Metaphysics as a Guide to Morals*. New York: Penguin Books, 1994.

———. *Existentialists and Mystics: Writings on Philosophy and Literature*, edited by Peter J. Conradi, 287–96. New York: Penguin Books, 1999a.

———. "Art Is the Imitation of Nature." In *Existentialists and Mystics: Writings on Philosophy and Literature*, edited by Peter J. Conradi, 243–58. New York: Penguin Books, 1999b.

———. "Knowing the Void." In *Existentialists and Mystics: Writings on Philosophy and Literature*, edited by Peter J. Conradi, 157–60. New York: Penguin Books, 1999c.

———. "The Darkness of Practical Reason." In *Existentialists and Mystics: Writings on Philosophy and Literature*, edited by Peter J. Conradi, 193–202. New York: Penguin Books, 1999d.

———. "The Sublime and the Good." In *Existentialists and Mystics: Writings on Philosophy and Literature*, edited by Peter J. Conradi, 205–20. New York: Penguin Books, 1999e.

———. "Vision and Choice in Morality." *In Existentialists and Mystics: Writings on Philosophy and Literature*, edited by Peter J. Conradi, 76–98. New York: Penguin Books, 1999f.

———. "On 'God' and 'Good'." In *Existentialists and Mystics: Writings on Philosophy and Literature*, edited by Peter J. Conradi, 337–362. New York: Penguin Books, 1999g.

———. *The Sovereignty of Good.* New York: Routledge, 2001.

Murphy, Jeffrie G., and Jean Hampton. *Forgiveness and Mercy.* New York: Cambridge University Press, 1988.

Nandy, Ashis. *The Intimate Enemy: Loss and Recovery of Self Under Colonialism.* New York: Oxford University Press, 2010.

Nash, William P., and Dewleen G. Baker. "Competing and Complementary Models of Combat Stress Injury." In *Combat Stress Injury: Theory, Research, and Management*, edited by Charles R. Figley and William P. Nash, 65–96. New York: Routledge, 2006.

Nash, William P., Teresa L. Marino Carper, Mary Alice Mills, Teresa Au, Abigail Goldsmith, and Brett T. Litz. "Psychometric Evaluation of the Moral Injury Events Scale." *Military Medicine* 178 (2013): 646–52.

Niebuhr, H. Richard. *The Responsible Self: An Essay in Christian Moral Philosophy.* Louisville, KY: Westminster John Knox, 1999.

Niebuhr, Reinhold. *An Interpretation of Christian Ethics.* Louisville, KY: Westminster John Knox Press, 2013.

Nussbaum, Martha C. *Love's Knowledge: Essays on Philosophy and Literature.* New York: Oxford University Press, 1992.

———. "Compassion: The Basic Social Emotion." In *The Communitarian Challenge to Liberalism*, edited by Ellen Frankel Paul, Fred D Miller, and Jeffrey Paul, 27–58. New York: Cambridge University Press, 1996a.

———. *The Therapy of Desire: Theory and Practice in Hellenistic Ethics.* Princeton, N.J.: Princeton University Press, 1996b.

———. *The Therapy of Desire: Theory.* "Virtue Ethics: A Misleading Category?" *The Journal of Ethics* 3 (1999): 163–201.

———. *The Fragility of Goodness: Luck and Ethics in Greek Tragedy and Philosophy.* New York: Cambridge University Press, 2001a.

———. *Women and Human Development: The Capabilities Approach.* New York: Cambridge University Press, 2001b.

———. *Frontiers of Justice: Disability, Nationality, Species Membership.* Cambridge, Mass: Belknap Press, 2007.

Oksenburg Rorty, Amélie. "The Vanishing Subject: The Many Faces of Subjectivity." In *Subjectivity: Ethnographic Investigations*, edited by Joao Biehl, Byron Good, and Arthur Kleinman, 34–51. Berkeley: University of California Press, 2007.

Ortner, Sherry B. "Theory in Anthropology since the Sixties." *Comparative Studies in Society and History* 26:1 (1984): 126–66.

———. "Subjectivity and Cultural Critique." *Anthropological Theory* 5:1 (March 2005): 31–52.

———. *Anthropology and Social Theory: Culture, Power, and the Acting Subject.* Durham, NC: Duke University Press, 2006.

Papanikolaou, Aristotle. "Trinity, Virtue, and Violence." In *God and the Moral Life*, edited by Myriam Renaud and Joshua Daniel, 115–34. New York: Routledge, 2017a.

———. "The Ascetics of War: The Undoing and Redoing of Virtue." In *Orthodox Christian Perspectives on War*, edited by Perry T. Hamalis and Valerie A. Karras, 13–36. Notre Dame, Ind.: University of Notre Dame Press, 2017b.

Patterson, Orlando. *Slavery and Social Death: A Comparative Study.* Cambridge, Mass: Harvard University Press, 1982.

Paul, Lisa A., Daniel F. Gros, Martha Strachan, Glenna Worsham, Edna B. Foa, and Ron Acierno. "Prolonged Exposure for Guilt and Shame in a Veteran of Operation Iraqi Freedom." *American Journal of Psychotherapy* 68 (2014): 277–86.

Perica, Vjekoslav. *Balkan Idols: Religion and Nationalism in Yugoslav States.* New York: Oxford University Press, 2004.

Plato. *The Laws of Plato.* Translated by Thomas L. Pangle. Chicago: University of Chicago Press, 1988.

———. *The Republic of Plato.* Translated by Allan Bloom. New York: Basic Books, 1991.

Power, Samantha. *A Problem from Hell: America and the Age of Genocide.* New York: Basic Books, 2013.

Prstojevic, Miroslav. *Sarajevo: Survival Guide.* Translated by Aleksandra Wagner. New York: Workman, 1994.

Ramet, Sabrina Petra. *Social Currents in Eastern Europe: Sources and Meaning of the Great Transformation.* Durham, NC: Duke University Press, 1991.

———. *Balkan Babel: The Disintegration of Yugoslavia From the Death of Tito to the Fall of Milosevic.* Boulder, CO: Westview Press, 2002.

Rawls, John. *A Theory of Justice.* Cambridge, Mass: Belknap Press, 1999.

Reitan, Richard M. *Making a Moral Society: Ethics and the State in Meiji Japan.* Honolulu: University of Hawaii Press, 2010.

Rieff, David. *Slaughterhouse: Bosnia and the Failure of the West.* New York: Touchstone, 1996.

Rilke, Rainer Maria. *Duino Elegies.* Falls Church, VA: Azul Editions, 1981.

Robbins, Joel. *Christianity and Moral Torment in Papua New Guinea Society.* Berkeley: University of California, 2004.

———. "Between Reproduction and Freedom: Morality, Value, and Radical Cultural Change." *Ethnos* 72:3 (2007): 293–314.

Robjant, David. "As a Buddhist Christian: The Misappropriation of Iris Murdoch." *The Heythrop Journal* 52 (2011): 993–1008.

———. "Symposium on Iris Murdoch: How Miserable We Are, How Wicked: Into the 'Void' with Murdoch, Mulhall, and Antonaccio." *The Heythrop Journal* 54:6 (November 2013): 999–1006.

Rorty, Richard. *Contingency, Irony, and Solidarity.* New York: Cambridge University Press, 1989.

Sayer, Andrew. *Why Things Matter to People: Social Science, Values and Ethical Life.* New York: Cambridge University Press, 2011.

Scarry, Elaine. *The Body in Pain: The Making and Unmaking of the World.* New York: Oxford University Press, 1987.

Scharen, Christian, and Aana Marie Vigen, eds. *Ethnography as Christian Theology and Ethics.* New York: Continuum, 2011.

Scheper-Hughes, Nancy. *Death Without Weeping: The Violence of Everyday Life in Brazil.* Berkeley: University of California Press, 1993.

———. "The Primacy of the Ethical: Propositions for a Militant Anthropology." *Current Anthropology* 36:3 (June 1995): 409–40.

Scheper-Hughes, Nancy and Philippe Bourgois, eds. *Violence in War and Peace: An Anthology.* Malden, MA: Blackwell, 2004.

Schmidt, Bettina, and Ingo Schroeder, eds. *Anthropology of Violence and Conflict.* New York: Routledge, 2001.

Sells, Michael. *The Bridge Betrayed: Religion and Genocide in Bosnia.* Berkeley: University of California Press, 1996.

Shay, Jonathan. "Casualties." *Daedalus* 140:3 (July 2011): 179–88.

Sherman, Nancy. "Recovering Lost Goodness: Shame, Guilt, and Self-Empathy." *Psychoanalytic Psychology* 31:2 (2014): 217–35.

———. *Afterwar: Healing the Moral Wounds of Our Soldiers.* New York: Oxford University Press, 2015.

Slingerland, Edward. "The Situationist Critique and Early Confucian Virtue Ethics." *Ethics* 121:2 (2001): 390–419.

Slote, Michael. *Morals from Motives.* New York: Oxford University Press, 2003.

Smith, Andrea. "Heteropatriarchy and the Three Pillars of Settler Colonialism." In *The Color of Violence: The INCITE! Anthology*, edited by Andrea Lee Smith, Beth E. Richie, Julia Sudbury, and Janelle White, 68–73. Boston: South End Press, 2006.

Smith, Erin R., Jeanne M. Duax, and Sheila A.M. Rauch. "Perceived Perpetration During Traumatic Events: Clinical Suggestions from Experts in Prolonged Exposure Therapy." *Cognitive and Behavioral Practice* 20 (2013): 461–70.

Smith, Jonathan Z. *Imagining Religion: From Babylon to Jonestown.* Chicago: University of Chicago Press, 1982.

Steenkamp, Maria M., William P. Nash, Leslie Lebowitz, and Brett T. Litz. "How Best to Treat Deployment-Related Guilt and Shame: Commentary on Smith, Duax, and Rauch." *Cognitive and Behavioral Practice* 20 (2013): 471–75.

Stein, Nathan R, Mary Alice Mills, Kimberly Arditte Hall, and Brett T. Litz. "A Scheme for Categorizing Behavior Modification." *Behavior Modification* 36 (2012): 787–807.

Stout, Jeffrey. "Commitments and Traditions in the Study of Religious Ethics." *The Journal of Religious Ethics* 25:3 (1997): 23–56.

Swanton, Christine. "Outline of a Nietzschean Virtue Ethics." *International Studies in Philosophy* 30:3 (1998): 23–38.

———. *Virtue Ethics: A Pluralistic View.* New York: Oxford University Press, 2005.

———. "A Challenge to Intellectual Virtue from Moral Virtue: The Case of Universal Love." *Metaphilosophy* 41:1–2 (January 2010): 152–71.

Tambiah, Stanley J. *Leveling Crowds: Ethnonationalist Conflicts and Collective Violence in South Asia.* Berkeley: University of California Press, 1997.

Tavory, Iddo. "The Question of Moral Action: A Formalist Position." *Sociological Theory* 29:4 (2011): 272–93.

Taylor, Charles. *Sources of the Self: The Making of the Modern Identity.* Cambridge, Mass: Harvard University Press, 1992.

———. *Modern Social Imaginaries.* Durham, NC: Duke University Press, 2003.

Tessman, Lisa. *Burdened Virtues: Virtue Ethics for Liberatory Struggles.* New York: Oxford University Press, 2005.

Tipton, Steven M. "Social Differentiation and Moral Pluralism." In *Meaning and Modernity: Religion, Polity, and Self,* edited by Richard Madsen, 15–40. Berkeley: University of California Press, 2002.

Trimble, Michael R. "Post-Traumatic Stress Disorder: History of a Concept." In *Trauma and Its Wake: The Study and Treatment of Post-Traumatic Stress Disorder,* edited by Charles R. Figley, 5–14. Bristol, PA: Brunner/Mazel, 1985.

Vargas, Alison Flipse, Thomas Hanson, Douglas Kraus, Kent Drescher, and David Foy. "Moral Injury Themes in Combat Veterans' Narrative Responses from the National Vietnam Veterans' Readjustment Study." *Traumatology* 19 (2013): 243–50.

Vazquez Torres, Jessica. "Does Moral Injury Have a Color? On Moral Injury and Race in the United States." Paper presented at the annual meeting of the American Academy of Religion, San Diego, CA, 2014.

Vrcan, Srdjan. "The Religious Factor and the War in Bosnia and Herzegovina." In *Religion and the War in Bosnia,* edited by Paul Mojzes, 108–31. Atlanta: Scholars Press, 1998.

Vulliamy, Ed. *Seasons in Hell: Understanding Bosnia's War.* New York: St. Martin's Press, 1994.

Walker, Rebecca L., and Philip J. Ivanhoe, eds. *Working Virtue: Virtue Ethics and Contemporary Moral Problems*. New York: Oxford University Press, 2009.

Waller, James. *Becoming Evil: How Ordinary People Commit Genocide and Mass Killing*. New York: Oxford University Press, 2002.

Weber, Max. "The Social Psychology of World Religions." In *From Max Weber: Essays in Sociology*, edited by H.H. Gerth and C. Wright Mills, 267–301. New York: Oxford University Press, 1946.

———. *The Sociology of Religion*. Translated by Ephraim Fischoff. Boston: Beacon Press, 1993.

———. *The Protestant Ethic and the Spirit of Capitalism*. Translated by Talcott Parsons. New York: Routledge, 2001.

Weil, Simone. *Waiting for God*. New York: HarperCollins, 2009.

Welch, Sharon. *A Feminist Ethic of Risk*. Minneapolis, MN: Fortress Press, 1990.

Wesselingh, Isabelle, and Arnaud Vaulerin. *Raw Memory: Prijedor, Laboratory of Ethnic Cleansing*. London: Saqi Books, 2005.

Widdows, Heather. *The Moral Vision of Iris Murdoch*. Burlington, VT: Ashgate, 2005.

Widlock, Thomas. "Virtue." In *A Companion to Moral Anthropology*, edited by Didier Fassin, 186–203. Malden, Mass: Wiley-Blackwell, 2012.

———. "Sharing by Default? Outline of an Anthropology of Virtue." *Anthropological Theory* 4:1 (March 2004): 53–70.

Williams, Bernard. "Moral Luck." In *Moral Luck: Philosophical Papers, 1973–1980*, 20–39. New York: Cambridge University Press, 1981.

———. *Ethics and the Limits of Philosophy*. Cambridge, Mass: Harvard University Press, 1986.

Wiinikka-Lydon, Joseph. "Moral Injury as Inherent Political Critique: The Prophetic Possibilities of a New Term." *Political Theology* 18:3 (2017): 219–232.

———. "Mapping Moral Injury: Comparing Discourses of Moral Harm." *The Journal of Medicine and Philosophy* 44:2 (2019): 175–91.

Worthen, Miranda, and Jennifer Ahern. "The Causes, Course, and Consequences of Anger Problems in Veterans Returning to Civilian Life." *Journal of Loss and Trauma* 19 (2014): 355–63.

Yu, Jiyuan. *The Ethics of Confucius and Aristotle: Mirrors of Virtue*. New York: Routledge, 2007.

Zagzebski, Linda. *Virtues of the Mind: An Inquiry into the Nature of Virtue and the Ethical Foundations of Knowledge*. New York: Cambridge University Press, 1996.

Zigon, Jarrett. "Moral Breakdown and the Ethical Demand: A Theoretical Framework for an Anthropology of Moralities." *Anthropological Theory* 7:2 (June 2007): 131–50.

———. *Morality: An Anthropological Perspective*. New York: Berg, 2008.

———. "On Love: Remaking Moral Subjectivity in Postrehabilitation Russia." *American Ethnologist* 40:1 (2013): 201–15.

INDEX[1]

[1] Note: Page numbers followed by 'n' refer to notes.

© The Author(s) 2019
J. Wiinikka-Lydon, *Moral Injury and the Promise of Virtue*,
https://doi.org/10.1007/978-3-030-32934-1

The manufacturer's authorised representative in the EU is Springer
Nature Customer Service Centre GmbH, Europaplatz 3, 69115 Heidelberg,
Germany. If you have any concerns regarding our products, please
contact ProductSafety@springernature.com

Printed and bound by CPI Group (UK) Ltd, Croydon, CR0 4YY

23/04/2026

02095601-0003